QUARTIC SURFACES

WITH SINGULAR POINTS

QUARTIC SURFACES

WITH SINGULAR POINTS

BY

C. M. JESSOP, M.A.

FORMERLY FELLOW OF CLARE COLLEGE, CAMBRIDGE
PROFESSOR OF MATHEMATICS IN ARMSTRONG COLLEGE
IN THE UNIVERSITY OF DURHAM

Cambridge:
at the University Press
1916

CAMBRIDGE
UNIVERSITY PRESS

University Printing House, Cambridge CB2 8BS, United Kingdom

Cambridge University Press is part of the University of Cambridge.

It furthers the University's mission by disseminating knowledge in the pursuit of education, learning and research at the highest international levels of excellence.

www.cambridge.org
Information on this title: www.cambridge.org/9781316601815

First published 1916
First paperback edition 2015

A catalogue record for this publication is available from the British Library

ISBN 978-1-316-60181-5 Paperback

PREFACE

THE purpose of the present treatise is to give a brief account of the leading properties, at present known, of quartic surfaces which possess nodes or nodal curves.

A surface which would naturally take a prominent position in such a book is the Kummer surface, together with its special forms, the tetrahedroid and the wave surface, but the admirable work written by the late R. W. H. T. Hudson, entitled *Kummer's Quartic Surface*, renders unnecessary the inclusion of this subject. Ruled quartic surfaces have also been omitted.

For the convenience of readers, a brief summary of all the leading results discussed in this book has been prefixed in the form of an Introduction.

I have to express my great obligation to Prof. H. F. Baker, Sc.D., F.R.S., who has given much encouragement and valuable criticism. Finally I feel greatly indebted to the staff of the University Press for the way in which the printing has been carried out.

C. M. JESSOP.

March, 1916.

CONTENTS

ART.		PAGE
	INTRODUCTION	xi

CHAPTER I.

QUARTIC SURFACES WITH ISOLATED SINGULAR POINTS.

ART.		
1, 2	Quartic surfaces with four to seven nodes	1
3	Quartic surfaces with more than seven nodes . . .	3
4, 5	Quartic surfaces with nodal sextic curves	4
6—12	Quartic surfaces with eight to sixteen nodes . . .	9

CHAPTER II.

DESMIC SURFACES.

13	Desmic tetrahedra	24
14	Quartic curves on the desmic surface	26
15—18	Expression of coordinates in terms of σ functions and resulting properties	27
19	Section by tangent plane	35
20	A mode of origin of the surface	35
21	The sixteen conics of the surface	37

CHAPTER III.

QUARTIC SURFACES WITH A DOUBLE CONIC.

22	Five cones touching surface along twisted quartic whose tangent planes meet surface in pairs of conics . .	38
23—28	Expression of coordinates in terms of two parameters and mapping of surface on a plane	40
29	The class of the surface	46
30	The sixteen lines of the surface	46
31	Origin of the surface by aid of two quadrics and a point.	47
32, 33	Perspective relationship with general cubic surface . .	49
34	Connection with plane quartic curve	52
35—37	Segre's method of projection in four-dimensional space .	55

CHAPTER IV.

QUARTIC SURFACES WITH A NODAL CONIC AND ADDITIONAL NODES.

ART.		PAGE
38	Case of one to four nodes	62
39—43	Origin and nature of double points arising from special positions of base-points	64
44—46	Cuspidal double curve	69
47	Double conic consists of a pair of lines	72
48—53	Segre's classification of different species of the surface	72

CHAPTER V.

THE CYCLIDE.

54, 55	Equation and mode of origin of cyclide	86
56	Inverse points on the surface	89
57—59	The fundamental quintic and form of the cyclide	90
60—62	Power of two spheres: pentaspherical coordinates	94
63—69	Canonical forms of the equation of the cyclide	99
70	Tangent and bitangent spheres of surface	108
71, 72	Confocal cyclides	109
73, 74	Sphero-conics on the surface	113
75	Cartesian equation of confocal cyclides	115
76	Common tangent planes of cyclide and tangent quadric	117

CHAPTER VI.

SURFACES WITH A DOUBLE LINE: PLÜCKER'S SURFACE.

77, 78	Conics on the surface	118
79—81	Mapping of the surface on a plane	120
82	Nodes on surface with a double line	125
83	Plücker's surface	128

CHAPTER VII.

QUARTIC SURFACES WITH AN INFINITE NUMBER OF CONICS: STEINER'S SURFACE: THE QUARTIC MONOID.

84, 85	Quartic surfaces with an infinite number of conics	131
86—88	Rationalization of Steiner's surface: all its tangent planes contain two conics of the surface: images of its asymptotic lines	134
89	Modes of origin of the surface	139
90	Quartic curves on the surface	141
91	Weierstrass's method of origin of the surface	143
92	Eckhardt's point-transformation	145
93	The quartic monoid	147

CHAPTER VIII.

THE GENERAL THEORY OF RATIONAL QUARTIC SURFACES.

ART.		PAGE
94	The cubics which rationalize $\sqrt{\Omega(y)}$ when $\Omega(y)=0$ is the general plane quartic: the surface $S_4^{(1)}$	152
95, 96	The curves which rationalize $\sqrt{\Omega(y)}$ when $\Omega(y)=0$ is respectively a sextic with a quadruple point or a sextic with two consecutive triple points: the surfaces $S_4^{(2)}$ and $S_4^{(3)}$	155

CHAPTER IX.

DETERMINANT SURFACES.

97	Sextic curves on the surface	161
98, 99	Correspondence of points on the surface. . . .	163
100	The Jacobian of four quadrics and the symmetroid .	165
101	Distinctive property of the symmetroid	167
102	Tangent plane of Jacobian	168
103, 104	Cubic, quartic, and sextic curves on the surfaces . .	169
105	Additional nodes on the surfaces	172
106—115	Weddle's surface	173
116	Bauer's surfaces.	189
117, 118	Schur's surfaces.	191
	INDEX OF SUBJECTS	196
	INDEX OF AUTHORS	198

ADDENDA

Throughout, the vertices of the tetrahedron of reference are denoted by A_1, A_2, A_3, A_4: see p. 50.

pp. 38, 45. The ∞^1 quadrics $\psi+2\lambda\phi+\lambda^2w^2=0$ touch the surface $\phi^2=w^2\psi$ along quadri-quartics. They are the quadrics mentioned on p. 59.

CORRIGENDA

p. 38, line 6, *for* $4w^2\psi$ *read* $w^2\psi$.
line 9, *for* close-points *read* pinch-points.

p. 40, last line but one, *for* be *read* be taken to be.
omit foot-note.

p. 76, foot-note, *insert* fourth edition.

INTRODUCTION

Ch. I. Quartic surfaces with isolated singular points.

This chapter, which is based on the results of Cayley* and Rohn, gives a method of classification of quartic surfaces which possess a definite number of isolated nodes and no nodal curves. The number of such nodes cannot exceed sixteen. Rohn has given a mode of classification for the surfaces having more than seven nodes, based on the properties of a type of seven-nodal plane sextic curves.

The equation of a quartic surface which has a node at the point $x = y = z = 0$, will be of the form

$$u_2 w^2 + 2u_3 w + u_4 = 0,$$

where $u_2 = 0$, $u_3 = 0$, $u_4 = 0$ are cones whose vertex is this point.

The tangent cone to the quartic whose vertex is the point is therefore

$$u_2 u_4 - u_3^2 = 0.$$

The section of this cone by any plane gives a plane sextic curve having a *contact-conic* u_2, i.e. a conic which touches the sextic where it meets it. When the surface has eight nodes the tangent cone whose vertex is any one of them will have seven double edges which give seven nodes on the plane sextic.

Such sextics are divided into two classes, viz. those for which there is an infinite number of cubics through the seven nodes, and two other points of the curve, and those for which there is only one such cubic. When a quartic surface is such that it has eight nodes consisting of the common points of three quadrics, the tangent cone from any node to the surface gives rise to a plane sextic of the first kind: such a quartic surface is said to be

* *Recent researches*, etc., Proc. Lond. Math. Soc. (1869–71

syzygetic: the equation of the surface is represented by an equation of the form

$$(a\between A, B, C)^2 = 0,$$

where $A=0$, $B=0$, $C=0$ represent quadrics whose intersections give the eight nodes.

The second, or general kind of sextic, arises from the general type of eight-nodal quartic surface which is said to be *asyzygetic*.

Similarly in the case of nine-nodal and ten-nodal quartic surfaces we have two kinds of plane sextics distinguished as above, giving rise to syzygetic and asyzygetic surfaces.

For ten-nodal surfaces there are two varieties of asyzygetic surfaces, one of which, the *symmetroid* (see Ch. IX), arises when the sextic curve consists of two cubic curves. The tangent cone from *each* of the ten nodes of this surface then consists of two cubic cones. There are also two varieties of ten-nodal syzygetic surfaces.

Seven points may be taken arbitrarily as nodes of a quartic surface, but if there is an eighth node it must either be the eighth point of intersection of the quadrics through the seven points, or, in the case of the general surface, lie upon a certain sextic surface, the *dianodal* surface, determined by the first seven nodes; hence it may not be taken arbitrarily.

When an eight-nodal surface has a ninth node the latter must lie on a curve of the eighteenth order, the *dianodal* curve.

Plane sextics with ten nodes and a contact-conic are divided into three classes according as they are the projections of the intersection of a quadric with (1) a cubic surface, (2) a quartic surface which also contains two generators of the same set of the quadric, (3) a quintic surface which also contains four generators of the same set of the quadric.

The first and second types of sextics are connected with eleven-nodal surfaces which are respectively asyzygetic and syzygetic; the third type gives a symmetroid with eleven nodes. A fourth surface arises when the sextic breaks up into two lines and a nodal quartic.

Twelve nodes on the quartic surface give rise to eleven nodes on the sextic, which must therefore break up into simpler curves; this process of decomposition goes on until we arrive at six straight lines, which case corresponds to the sixteen-nodal or Kummer surface.

There are four varieties of surfaces with twelve nodes of which one is a symmetroid: there are only *two* varieties of surfaces with thirteen nodes and only *one* with fourteen nodes, viz. that given by the equation

$$\sqrt{xx'} + \sqrt{yy'} + \sqrt{zz'} = 0.$$

An additional node arises for a surface having this equation, when there exists between the planes $x \ldots z'$ the identity

$$Ax + By + Cz + A'x' + B'y' + C'z' \equiv 0,$$

with the condition

$$AA' = BB' = CC'.$$

If another such relation exists between the planes $x \ldots z'$, there is a sixteenth node.

Ch. II. Desmic surfaces.

A surface of special interest which possesses nodes and no singular curve is the desmic surface. Three tetrahedra $\Delta_1, \Delta_2, \Delta_3$ are said to form a desmic system when an identity exists of the form

$$\alpha\Delta_1 + \beta\Delta_2 + \gamma\Delta_3 \equiv 0,$$

where Δ_i is the product of four factors linear in the coordinates.

It is easily deducible from this identity that the tetrahedra are so related that every face of Δ_3 passes through the intersection of faces of Δ_1 and Δ_2; hence we have sixteen lines through each of which one face of each tetrahedron passes. It is deducible as a consequence, that any pair of opposite edges of Δ_1 together with a pair of opposite edges of Δ_2 form a skew quadrilateral; and so for Δ_1 and Δ_3, Δ_2 and Δ_3.

It also follows that if the edges A_1A_2, A_1A_3, A_1A_4 of Δ_1 meet the respective edges of Δ_2 in LL', MM', NN'; then A_1, L, A_2, L' are four harmonic points; and so for A_1MA_3M', A_1NA_3N'. The relationship between the three tetrahedra is entirely symmetrical.

Hence we may construct a tetrahedron desmic to a given tetrahedron Δ, by drawing through any point A the three lines which meet the three pairs of opposite edges of Δ, then if the intersections of these three lines with the edges of Δ be LL', MM', NN' respectively, the fourth harmonics to A, L, L'; A, M, M'; A, N, N' will, with A, form a tetrahedron desmic to Δ.

The join of any vertex of Δ_1 and any vertex of Δ_2 passes through a vertex of Δ_3: there are therefore sixteen lines upon each of which one vertex of each tetrahedron lies. Hence any two desmic tetrahedra have four centres of perspective, viz. the vertices of the third tetrahedron.

If Δ_1 be taken as tetrahedron of reference the identity connecting Δ_1, Δ_2, Δ_3 is given by the equation

$$\begin{aligned}16xyzt &\\ &-(x+y+z+t)(x+y-z-t)(x-y+z-t)(x-y-z+t)\\ &-(x+y+z-t)(x+y-z+t)(x-y+z+t)(-x+y+z+t)=0.\end{aligned}$$

Closely connected with the system of tetrahedra Δ_i is a second desmic system of three tetrahedra D_i. They are afforded by the identity

$$(x^2-y^2)(z^2-t^2)+(x^2-t^2)(y^2-z^2)+(x^2-z^2)(t^2-y^2)=0.$$

The sixteen lines joining the vertices of the Δ_i are the sixteen intersections of the faces of the D_i.

A desmic surface is such that a pencil of such surfaces contains each of three such tetrahedra D_i in desmic position. The surface has as nodes the vertices of the corresponding tetrahedra Δ_i; hence the sixteen lines joining the vertices of the latter tetrahedra lie on the surface: along each of them the tangent plane to the surface is the *same*, i.e. the line is torsal; the tangent plane meets the surface also in a conic, and hence there are sixteen conics on the surface lying in these tangent planes.

There is a doubly-infinite number of quadrics through the vertices of any two tetrahedra Δ_i, the surface is therefore syzygetic; these quadrics meet the surface in three singly-infinite sets of quadri-quartics; one curve of each set passes through any point of the surface.

The coordinates of any point on the surface can be expressed in terms of two variables u, v as follows:

$$\rho x=\frac{\sigma_1(u)}{\sigma_1(v)},\quad \rho y=\frac{\sigma_2(u)}{\sigma_2(v)},\quad \rho z=\frac{\sigma_3(u)}{\sigma_3(v)},\quad \rho t=\frac{\sigma(u)}{\sigma(v)};$$

since this leads to

$$\begin{aligned}(e_1-e_2)(x^2y^2+z^2t^2)+(e_3-e_1)(x^2z^2+y^2t^2)&\\ +(e_2-e_3)(x^2t^2+y^2z^2)&=0,\end{aligned}$$

which is one form of equation belonging to the surface.

The three systems of twisted quartics are obtained by writing respectively

$$v = \text{constant}, \quad u - v = \text{constant}, \quad u + v = \text{constant}.$$

The generators of the preceding doubly-infinite set of quadrics form a cubic complex which depends merely on the twelve desmic points; all the lines through these points belong to the complex. Any line of this complex meets the surface in points whose arguments (u, v) are respectively

$$(\beta + \mu, \alpha), \quad (\beta - \mu, \alpha), \quad (\alpha + \mu, \beta), \quad (\alpha - \mu, \beta).$$

The tangents to the three quadri-quartics which pass through any point of the surface are bitangents of the surface, and their three other points of contact are collinear.

The curves $u = \text{constant}$, $v = \text{constant}$ form a conjugate system of curves on the surface: the system conjugate to $u + v = \text{constant}$ is $3v - u = \text{constant}$; the system conjugate to $u - v = \text{constant}$ is $3v + u = \text{constant}$; hence we derive the differential equation of conjugate tangents as

$$du\,du_1 + 3dv\,dv_1 = 0.$$

The points of any plane section of the surface are divided into sets of sixteen points, lying upon three sets of four lines belonging to the cubic complex, where each line contains four of the sixteen points; denoting these twelve lines by $a_1 \ldots a_4$, $b_1 \ldots b_4$, $c_1 \ldots c_4$, then if C is the curve enveloped by the lines of the cubic complex in the plane, the points of contact of the lines a lie on a tangent α of C, those of the lines b on a tangent β, and those of the lines c on a tangent γ; where α, β, γ are three concurrent lines.

If p, q, r are three lines of a cubic surface forming a triangle, then any three planes through p, q, r respectively meet the cubic surface in conics which lie on the same quadric; the locus of the vertices of such of these quadrics as are cones is a desmic surface.

Ch. III. Quartic surfaces with a double conic.

The equation of a quartic surface with a nodal conic has the form

$$\phi^2 = w^2\psi.$$

This may be written

$$(\phi + \lambda w^2)^2 = w^2 (\psi + 2\lambda\phi + \lambda^2 w^2);$$

and hence can be brought to the form

$$\Phi^2 = w^2 V,$$

where $V = 0$ is a quadric cone, in five ways. Each tangent plane of the cones V_i meets the surface in a pair of conics. Among the conics arising from any particular cone V_1 there are eight pairs of lines; hence the surface contains sixteen lines. The relationship of these lines as regards intersection is the same as that of sixteen lines of the general cubic surface obtained by omitting any of its twenty-seven lines, p, together with the ten lines which intersect p.

The coordinates of any point on the surface can be expressed as cubic functions of two parameters by the equations

$$\rho x_i = f_i(\xi_1, \xi_2, \xi_3), \qquad (i = 1, 2, 3, 4);$$

so that every plane section of the surface is represented by a member of the family of curves $\sum_1^4 \alpha_i f_i = 0$; where $f_1 = 0, \ldots, f_4 = 0$ are plane cubic curves which have five common points; hence the surface is rational and is represented on a plane. Each of these five points, the *base-points* of the representation, is the *image* of a line of the surface. The other lines of the surface are represented in the plane by the conic through the base-points and by the ten lines joining pairs of base-points.

This method enables us to determine the varieties of curves of different orders which can exist on the surface, by use of the equation

$$N = 3n - \Sigma \alpha_i,$$

where N is the order of the curve on the surface, n that of its image in the plane, and α_i the number of times the curve on the surface meets one of the lines represented by the base-points. It is found that the sixteen lines previously mentioned are the only lines on the surface; the only conics on the surface, apart from the double conic, are those in the tangent planes of the cones V_i.

We obtain ∞^2 twisted cubics on the surface, and also ∞^4 quadri-quartics together with ∞^3 twisted quartics of the second species. It is seen that the quadrics

$$\psi + 2\lambda\phi + \lambda^2 w^2 = 0$$

touch the surface along quartics. The class of the surface is twelve.

The surface may also be obtained by aid of any two given quadrics Q and H and any given point O, as follows: the surface is the locus of a point P such that the points O, P, K, P' are harmonic, P and P' conjugate for H, and K any point of Q; P' also lies on the surface.

The twenty-one constants of the surface are seen to arise from those of Q and H, and the coordinates of O. This point is the vertex of one of the five cones V_i; the vertices of the other four cones are the vertices of the tetrahedron which is self-polar for Q and H. The double conic is the intersection of H with its polar plane for O.

From the foregoing mode of origin of the surface O is said to be a *centre of self-inversion* of the surface with regard to the quadric H.

The surface may be related to the general cubic surface by a (1, 1) correspondence in two ways, the relationship being a perspective one in each case.

The surface is connected with the general quartic curve as follows: the tangent cone drawn to the surface from any point P of the double conic is of the fourth order, its section being the general quartic curve; the tangent planes from P to the five cones V_i, and the tangent planes to the surface at P, meet the plane of the quartic curve in lines bitangent to this curve.

The other sixteen bitangents arise from the planes passing through P and the sixteen lines of the surface. The cone whose vertex is P and base a conic of the surface meets the plane of the quartic curve in a conic which has four-point contact with the quartic.

The general quartic surface with a double conic is obtained by Segre as the projection from any point A of the intersection Γ of two quadratic manifolds or *varieties* $P=0, \Phi=0$, in four dimensions, upon any given hyperplane S_3. Among the varieties of the pencil $F+\lambda\Phi=0$ there are five *cones*, i.e. members of the pencil containing only four variables homogeneously; each cone possesses an infinite number of generating planes consisting of two sets, and each generating plane meets Γ in a conic. These generating planes are projected from A upon S as the tangent planes of a quadric cone. Hence arise the five cones of Kummer, and the conics lying in their tangent planes.

The double conic is obtained as the projection from A on S of the quadri-quartic which is the intersection with Γ of the tangent hyperplane at A of the variety which passes through A. When A lies on one of the five cones of the pencil $F+\lambda\Phi=0$, this quadri-quartic becomes two conics in planes whose line of intersection passes through A. Hence the conics are projected into intersecting *double lines* of the quartic surface. By this projective method the lines and conics of the quartic surface may be obtained, as also its properties generally.

Ch. IV. Quartic surfaces with a nodal conic and additional nodes.

A quartic surface with a nodal conic may also have isolated nodes, but their number cannot exceed four. Each such node is the vertex of a cone of Kummer, and for every node the number of these cones is reduced by unity. There are two kinds of surfaces with two nodes, in one case the line joining the nodes lies on the surface, and in the other case it does not. Nodes arise when the base-points of the representation of the surface on a plane have certain special positions; if either two base-points coincide, or if three are collinear, there is a node on the surface. If *either* a coincidence of two base-points *or* a collinearity of three base-points occurs twice, the quartic surface has two nodes and is of the first kind just mentioned; if there is one coincidence together with one collinearity, the quartic surface is of the second kind.

There are three nodes when two base-points coincide and also two of three collinear base-points coincide; finally, when the join of two coincident base-points meets the join of two other coincident base-points in the fifth base-point, there are four nodes.

Three coincident base-points give rise to a binode, *four* coincident base-points give rise to a binode of the second kind, i.e. when the line of intersection of the tangent planes lies in the surface, and *five* to a binode of the third species, i.e. when the line of intersection is a line of contact for one of the nodal planes.

When four base-points come into coincidence in an indeterminate manner we have a ruled surface; a special variety occurs when the fifth base-point coincides with them in a determinate manner.

The double conic may be *cuspidal,* i.e. when the two tangent planes to the surface at each point of it coincide; the class of this surface is six. The equation of the surface may in this case be reduced to the form

$$U^2 + x_1^3 x_2 = 0.$$

The surface has two *close-points* C, C' given by

$$x_1 = x_2 = U = 0.$$

If K be any point of CC' and π the polar plane of K for $U = 0$, then if any line through K meets π in L, it will meet the surface in four points P, P'; Q, Q' such that the four points K, P, L, P' and K, Q, L, Q' are harmonic.

The double conic may consist of two lines; the necessary condition for this is that three cubics of the system representing plane sections should be

$$\alpha u = 0, \quad \alpha v = 0, \quad \beta u = 0,$$

where $\alpha = 0$, $\beta = 0$ are lines, and $u = 0$, $v = 0$ are conics. Either or both of the double lines may be cuspidal.

Segre's method (Ch. III) affords a means of complete classification of quartic surfaces with a double conic, by aid of the theory of elementary factors. We thus obtain seven types, each type leading to sub-types.

There exists in the case of certain of these sub-types a *cone of the second order* in the pencil (F, Φ), i.e. a cone whose equation contains only three variables, say x_1, x_2, x_3; if the line $x_1 = x_2 = x_3$, which may be termed the *edge* of this cone, lies upon Γ, the surface is ruled. If the point of projection, A, is so chosen that the tangent hyperplane for A, of the variety which passes through A, is also a tangent hyperplane of this cone of the second order, the double conic is cuspidal.

When the pencil (F, Φ) consists entirely of cones of the first order having a common generator, and a common tangent hyperplane along this generator, the surface is that of Steiner.

Segre's table, which distinguishes each surface that can arise, is given on pp. 82–85.

Ch. V. The cyclide.

When the double conic is the section of a sphere by the plane at infinity, we obtain the cyclide. The equation of the cyclide is therefore $S^2 + U = 0$, where $S = 0$ is a sphere and $U = 0$ is a quadric.

This equation may be written in the form

$$\{x^2 + y^2 + z^2 - 2\lambda\}^2 + 4\{(A_1 + \lambda)x^2 + (A_2 + \lambda)y^2 + (A_3 + \lambda)z^2 + 2B_1x + 2B_2y + 2B_3z + C - \lambda^2\} = 0.$$

The second member of the left side will give a cone when λ is a root of the quintic $F(\lambda) = 0$, where $F(\lambda)$ is the discriminant of the second member. We thus obtain as in Ch. III five cones V_i; the tangent planes of each cone meet the surface in pairs of circles.

There are five sets of bitangent spheres of the surface; each sphere of any set cuts a fixed sphere orthogonally, and its centre lies on a fixed quadric. The centres of the five fixed spheres are the vertices of the cones V_i.

These five spheres $S_1 \ldots S_5$ are mutually orthogonal, and the centres of any four of them form a self-polar tetrahedron for the fifth sphere and its corresponding quadric Q.

The equations of a pair S_i, Q_i are respectively

$$x^2 + y^2 + z^2 + \frac{2B_1x}{A_1 + \lambda_i} + \frac{2B_2y}{A_2 + \lambda_i} + \frac{2B_3z}{A_3 + \lambda_i} + 2\lambda_i = 0;$$

$$\frac{x^2}{A_1 + \lambda_i} + \frac{y^2}{A_2 + \lambda_i} + \frac{z^2}{A_3 + \lambda_i} + 1 = 0;$$

where λ_i is one of the roots of $F(\lambda) = 0$.

The five quadrics $Q_1 \ldots Q_5$ are confocal; the curve of intersection of a pair S_i, Q_i is a focal curve of the surface.

The centre of a sphere S_i is a centre of self-inversion for the surface.

Three of the quadrics Q_i are necessarily real together with their corresponding spheres: one is an ellipsoid, one a hyperboloid of one sheet and one a hyperboloid of two sheets.

The surface is also obtained as the locus of the limiting points defined by S_i and the tangent planes of Q_i. Taking Q_i as an ellipsoid, this shows the shape of the surface to be one of the following:

(i) two ovals, one within the other, when S_i, Q_i do not intersect;

(ii) two ovals, external to each other, *or* a tubular surface similar to the anchor-ring, when the focal curve (S_i, Q_i) consists of two portions;

(iii) one oval, when the focal curve (S_i, Q_i) consists of one portion.

When $(\lambda + A_i)^2$ is a factor of $F(\lambda)$, one of the cones V is a pair of planes. If two roots of $F(\lambda)$ are equal, one of the principal spheres is a point-sphere. In a real cyclide only one principal sphere can be a point-sphere. Real cyclides must possess at least two principal spheres which are not point-spheres.

If $S_1 = 0, \dots S_5 = 0$ are any five spheres, there is a quadratic identity between the quantities $S_1 \dots S_5$, viz. that given by the equation

$$\begin{vmatrix} 0 & S_1 & \dots\dots\dots & S_5 \\ S_1 & -2r_1^2 & \pi_{12} \;\dots & \pi_{15} \\ S_2 & \pi_{12} & -2r_2^2 & \dots\dots \\ \cdot & \cdot & & \\ \cdot & \cdot & & \\ S_5 & \pi_{15} & \dots\dots\dots & -2r_5^2 \end{vmatrix} = 0,$$

where $r_1 \dots r_5$ are the radii of the spheres, and π_{ij} is the mutual *power* of the spheres $S_i = 0$, $S_j = 0$.

By solution of the equations

$$S_1 \equiv x^2 + y^2 + z^2 + 2f_1x + 2g_1y + 2h_1z + c_1, \text{ etc.},$$

it is seen that $x^2 + y^2 + z^2$, x, y, z, and unity, can be expressed as linear functions of $S_1 \dots S_5$; hence the equation of a cyclide appears as a quadratic function of $\frac{S_1}{r_1}, \dots \frac{S_5}{r_5}$, which are themselves connected by a quadratic identity. This gives rise to seven chief types of cyclide, by application of the theory of elementary factors; but only three of them give real cyclides, viz.

[11111], [2111], [311].

Each of these types and the corresponding sub-types, with the exception of the general cyclide, arise as the inverses of quadrics. The sub-type [(11) 111] can be expressed in terms of three variables. It is the envelope of spheres which pass through a fixed point and whose centres lie on a conic; contact with the envelope here occurs along a circle. It has also two systems of bitangent spheres, as in the general case. A variable sphere of one of these systems makes with two fixed spheres of the first system angles whose sum or whose difference is constant. The inverse of this cyclide is a cone.

The cyclide [(11) (11) 1] is known as *Dupin's cyclide*. There are two systems of spheres which touch the cyclide along circles; the spheres of each system cut one of the principal spheres at a constant angle. The spheres of either system are obtained as those which touch any two fixed spheres of the other system and have their centres on a given plane.

Denoting $\frac{S_i}{r_i}$ by x_i, the equation of the general cyclide appears as $\sum_1^5 a_i x_i^2 = 0$, with the condition $\sum_1^5 x_i^2 = 0$.

The system of cyclides $\sum_1^5 \frac{x_i^2}{a_i + \lambda} = 0$ is confocal with the first cyclide. Three confocals pass through any point and cut orthogonally.

The system of quadrics $V = 0$, where $V \equiv U + kS - k^2$, which touch the cyclide $S^2 + 4U = 0$ along sphero-conics are such that *two* of them pass through any point, *three* touch any line, *four* touch any plane. The four points of contact of the surfaces V which touch any given plane π are the centres of self-inversion for the section of the cyclide by π.

The locus of points of contact of common tangent planes of the cyclide and any given quadric V is a line of curvature on the cyclide.

The Cartesian equation of the system of confocals is

$$(A_1 + \lambda)(A_2 + \lambda)(A_3 + \lambda) S^2 + 4F(\lambda) Q = 0,$$

where S, Q have the same form as the S_i, Q_i when λ is substituted for λ_i.

The confocals to the given cyclide $S^2 + 4U = 0$, where

$$S \equiv x^2 + y^2 + z^2 - 2\lambda,$$

may be obtained as follows: when $S + 2L = 0$ is a point-sphere and $U + L^2 = 0$ is a cone, the locus of the centres of these point-spheres is a cyclide confocal with $S^2 + 4U = 0$.

Ch. VI. Surfaces with a double line: Plücker's surface.

The quartic surface with a double line is cut by any plane through the double line in a conic also. In eight cases this conic breaks up into a pair of lines, giving sixteen lines on the surface. There is no other line on the surface with the exception of the double line.

There are sixty-two planes not passing through the double line each of which meets the surface in a pair of conics, one of whose intersections lies on the double line. By aid of one of these conics c^2 the surface may be represented on a plane; for through any point x of the surface one line can be drawn to meet c^2 and also the double line, so that with each point of the surface one such line is associated. This line is determined as the intersection of two planes, each of whose coefficients contains linearly and homogeneously three parameters ξ_1, ξ_2, ξ_3. A third equation, arising from the equation of the surface, is that of a plane whose coefficients are quadratic in the ξ_i, and the intersection of these three planes is a point on the surface; hence we obtain a (1, 1) correspondence between the points x of the surface and the points ξ of a plane.

There are nine *base-points* in the plane, eight of which we represent by $B_1 \dots B_8$; they correspond to the points of eight non-intersecting lines of the surface, together with a point A which corresponds to any point of the conic coplanar with c^2. These nine points cannot constitute the complete intersection of two cubic curves.

To any plane section of the surface there corresponds, in the plane of ξ, a quartic curve having a node at A and passing through the points B_i. The cubic through the nine base-points corresponds to the double line.

The plane image of any curve of order M on the surface is a curve of order m, where

$$M = 4m - 2\beta - \Sigma\alpha,$$

β being the number of times the curve on the surface meets the conic corresponding to A, and $\Sigma\alpha$ the total number of passages of the image through the points B_i.

By applying Rohn's method to the surface, using any point on the double line as that from which a tangent cone is drawn, it is easy to see the modifications which arise when isolated nodes exist.

The section of the tangent cone, whose vertex is any point of the double line, is a sextic curve, meeting the double line in a quadruple point; with each additional node of the surface this curve acquires an additional node: when there are seven nodes the sextic becomes a nodal cubic, meeting the double line in one point together with three lines through this point. When the surface has eight nodes, the sextic curve becomes a conic together with four lines concurring at a point of the double line.

In the case of seven nodes there are three torsal lines meeting the double line, and each containing two nodes; also there are four tropes meeting in the seventh node, and each containing four nodes. If there are eight nodes we have Plücker's surface which has also eight tropes. The nodes form two tetrahedra, each of which is inscribed in the other. The nodes lie in pairs on four torsal lines meeting the double line. Through any two nodes not on the same torsal line there pass two tropes. The tropes can be arranged in four pairs so that the line of intersection of a pair meets the double line in a pinch-point.

Plane sections of Plücker's surface are represented by quartic curves having a common node and touching, at fixed points, four concurrent lines.

Ch. VII. Quartic surfaces containing an infinite number of conics: Steiner's surface: the quartic monoid.

The nature of the quartic surfaces which contain an infinite number of conics was investigated by Kummer. He showed the existence of the following classes: surfaces with a double conic or a double line; ruled quartic surfaces; the surface $\Phi^2 = \alpha\beta\gamma\delta$, where $\Phi = 0$ is a quadric and $\alpha, \beta, \gamma, \delta$ coaxal planes; Steiner's surface.

To these surfaces discussed by Kummer must be added the surface whose equation is

$$\{xw + f(y, z, w)\}^2 = (z, w \between a)^4.$$

The surface $\Phi^2 = \alpha\beta\gamma\delta$ has two tacnodes at the intersection of the common axis of the planes with Φ; it is birationally transformable into a cubic cone. The conics of the surface can be arranged in sets of four lying on the same quadric; the quadric cone whose vertex is on the axis of the planes $\alpha \ldots \delta$, and whose base is any conic of the surface, meets the surface in four conics.

Steiner's surface is of the third class and has four tropes; the coordinates of any point of the surface are expressible as homogeneous quadratic functions of three variables; conversely any surface, the coordinates of whose points are so expressible, is a Steiner surface. The surface has a triple point, three double lines meeting in the triple point, and a node on each double line. A characteristic property of the surface is that its section by any tangent plane breaks up into two conics. Every algebraic curve on the surface is of even order.

The surface being determined by the equations

$$\rho x_i = f_i(\eta_1, \eta_2, \eta_3) \qquad (i = 1, 2, 3, 4),$$

we are enabled to map the surface on the plane of the η_i.

Any conic of the surface is represented on the η-plane by a straight line, the pair of lines representing two conics in the same tangent plane of the surface are represented by the equations

$$\eta_1 + m\eta_2 + n\eta_3 = 0, \quad \eta_1 + \frac{1}{m}\eta_2 + \frac{1}{n}\eta_3 = 0.$$

The surface contains ∞^5 quartic curves of the second species, which are represented by the general conic in the plane of η; also ∞^4 quadri-quartics having a node on one of the double lines; they lie on quadrics passing through two double lines, and are represented by conics $\Sigma a_{ik}\eta_i\eta_k = 0$, in which two of the quantities a_{11}, a_{22}, a_{33} are equal.

The conics apolar to the four conics $f_i = 0$ form the pencil

$$u_\alpha^2 + \lambda u_\beta^2 = 0;$$

the conics of this pencil are inscribed in the same quadrilateral, and form the images on the plane of η of the asymptotic lines of the surface.

A form of the preceding property of the surface, that its coordinates are expressible as homogeneous quadratic functions of two variables, is the following: in the general quadric transformation

$$\rho x_i = f_i(\alpha_1, \alpha_2, \alpha_3, \alpha_4),$$

the locus of x is a Steiner surface when the locus of α is a plane. From this we derive the fact that Steiner's surface, and the cubic polar of a plane with reference to a general cubic surface, are reciprocal.

Another mode of origin of the surface, given by Sturm, is that if a pencil of surfaces of the second class is projectively related to the points of a line in such a way that the line meets one *conic* c^2 of the system in a point corresponding to c^2, and another *conic* c'^2 in a point corresponding to c'^2, then the envelope of the tangent cones drawn from the points of the line to the corresponding surfaces is a Steiner surface.

Weierstrass and Schröter have shown that a Steiner surface arises as a locus connected with a known theorem for the quadric. The theorem is that if through any given point A of a quadric any three mutually perpendicular lines are drawn, meeting the quadric again in L, M, N, then the plane LMN meets the normal at A in a fixed point.

This theorem may be generalized as follows: if A be joined to the vertices of *any* triangle self-polar for a given conic c^2 in a given plane α, and the joining lines meet the quadric again in L, M, N, then the plane LMN meets the line AR in a fixed point S, where R is the pole for c^2 of the trace on α of the tangent plane to the quadric at A.

If now c^2 is a member of the ∞^2 conics

$$\eta_1 U + \eta_2 V + \eta_3 W = 0,$$

where $U=0$, $V=0$, $W=0$ are given conics, we have a point S determined for each set of values of $\eta_1 : \eta_2 : \eta_3$. On giving these ratios all values the locus of S is a Steiner surface; for it can be shown that if the coordinates of S are $y_1 \ldots y_4$, we have

$$y_1 : y_2 : y_3 : y_4 = f_1(\eta) : f_2(\eta) : f_3(\eta) : f_4(\eta),$$

where the f_i are quadratic functions of the η_i.

Properties of Steiner's surface may be deduced by aid of the transformation

$$x_i y_i = \rho \qquad (i = 1, 2, 3, 4),$$

applied to any plane $\Sigma a_i x_i = 0$, giving the cubic surface

$$\Sigma \frac{a_i}{y_i} = 0,$$

which is the reciprocal of Steiner's surface.

Steiner's surface is one example of a type of surfaces known as *monoids*, viz. surfaces of the nth order which have an $(n-1)$-fold point. The equation of the quartic monoid may be written

$$wu_3 + u_4 = 0,$$

where $u_3 = 0$, $u_4 = 0$ are cones having their vertices at the triple point. The surface contains twelve lines, the intersections of $u_3 = 0$ and $u_4 = 0$. The surface is projectively related to any plane, e.g. the plane $w = 0$, in a (1, 1) manner, except that *every* point of each of these twelve lines is represented by *one* point only, viz. where the line meets the plane $w = 0$.

The surface contains conics in planes through two of the twelve lines, and twisted cubics on quadric cones passing through five of the twelve lines. The ∞^1 quadric cones passing through any four of the twelve lines meet the surface in quartic curves having a node at the triple point; the ∞^1 cubic cones passing through any eight of the twelve lines meet the surface in quartic curves without double points. If the lines corresponding to a curve of each type together make up the twelve lines, these two curves lie on one quadric. All these quartic curves are quadri-quartics.

Quartic curves of the second species arise as the intersection with the surface of cubic cones having six of the twelve lines as simple lines and one of them as double line; there are 5544 such quartic curves on the surface. The surface will have a line not passing through the triple point provided that three of the twelve lines are coplanar.

The cases of the quartic monoid of special interest are those in which there are six nodes; here the twelve lines coincide in pairs six times.

There are two cases of such surfaces; in the first case the six nodes may have any positions, this surface is a special case of the symmetroid; for the symmetroid being the result of eliminating the x_i from the equations

$$\alpha_1 \frac{\partial S_1}{\partial x_i} + \alpha_2 \frac{\partial S_2}{\partial x_i} + \alpha_3 \frac{\partial S_3}{\partial x_i} + \alpha_4 \frac{\partial S_4}{\partial x_i} = 0, \qquad (i = 1, 2, 3, 4),$$

where the α_i are regarded as point-coordinates, the surface considered is the special case in which one of the quadrics $S_i = 0$ is a plane taken doubly. The tangent cone to the surface whose vertex is one of the six nodes breaks up into two cubic cones. In the other case the six nodes lie on a conic whose plane is a trope of the surface. Each kind of surface has the same number of constants, viz. twenty-one.

Ch. VIII. Rational quartic surfaces.

The quartic surfaces with a triple point or with a double curve have been seen to be rational, i.e. the coordinates of the points of such a surface are expressible as rational functions of two parameters. Nöther has shown that there are only three rational quartic surfaces apart from them. The first of these surfaces has a *tacnode*, i.e. is such that every plane through the node meets the surface in a quartic curve having two consecutive double points at the node. The coordinates of any point x of the surface are projectively related to the points y of a *double plane* by the equations

$$\rho x_1 = y_1, \quad \rho x_2 = y_2, \quad \rho x_3 = y_3, \quad \rho x_4 = \frac{-\chi_2(y) \pm \sqrt{\Omega(y)}}{y_1},$$

where $\chi_2(y) = 0$ is a conic and $\Omega(y) = 0$ is a general quartic curve.

Clebsch showed that the points y can be expressed as rational functions of new variables z_i in such a way as to render $\sqrt{\Omega(y)}$ a rational function of the z_i, viz. by equations of the form

$$\sigma y_i = f_i(z), \qquad (i = 1, 2, 3),$$

where the curves $f_i(z) = 0$ are cubics having seven points in common. The plane sections of the surface have then as their images, in the field of the z_i, *sextic curves* having the seven points as nodes and also four other common points; the eleven points lie on the same cubic.

If the quartic surface has the equation

$$x_4^2 f_2 + 2x_4 f_3 + f_4 = 0,$$

we obtain

$$x_1 : x_2 : x_3 : x_4 = y_1 : y_2 : y_3 : \frac{-f_3(y) \pm \sqrt{\Omega(y)}}{f_2(y)},$$

where $\Omega(y) = 0$ is a sextic curve. It is shown that $\sqrt{\Omega(y)}$ is capable of rationalization only in the following two cases, viz. (1) when $\Omega(y) = 0$ is a sextic with a quadruple point; (2) when $\Omega(y) = 0$ is a sextic with two consecutive triple points.

The transformation to the simple plane is effected by the consideration that to plane sections through the double point there must correspond, in the simple plane, curves of order n of the same genus as these sections, viz. two, and intersecting each other in two variable points. This gives the equations

$$n^2 - 2 = \alpha_1 + 4\alpha_2 + \ldots + r^2\alpha_r,$$

$$\frac{n(n+3)}{2} - 1 = \alpha_1 + 3\alpha_2 + \ldots + \frac{r(r+1)}{2}\alpha_r,$$

where α_1 is the number of points the curves in the z-plane have in common, α_2 the number of double points they have in common, and so on. By aid of Cremona transformations repeatedly applied, it is seen that these curves of order n are capable of being replaced by one of the following types: (1) the curves $c_4(a^2b_1 \ldots b_{10})$, (2) the curves $c_6(a_1^2 \ldots a_8^2 b_1 b_2)$; i.e. quartic curves with one common node and ten common points, or sextic curves with eight common nodes and two common points. In each case the fixed points lie on one cubic.

The substitutions $\rho y = c_4(z)$, $\rho y = c_6(z)$ will then rationalize $\sqrt{\Omega(y)}$ in the two cases respectively mentioned, and hence lead to two rational quartic surfaces.

Ch. IX. Determinant surfaces.

The quartic surface whose equation is $\Delta = 0$, where Δ is a determinant of four rows whose elements are linear functions of the coordinates, depends upon thirty-three constants, one less than the general quartic surface. Taking as its equation

$$\begin{vmatrix} p_x & q_x & r_x & s_x \\ p_x' & q_x' & \cdots & \\ p_x'' & \cdots & & \\ p_x''' & \cdots & & \end{vmatrix} = 0,$$

it is seen that the surface contains two sets of sextic curves, viz. the curves

$$\left\Vert \begin{matrix} p_x & \cdots & & s_x \\ p_x' & \cdots & & \\ p_x'' & \cdots & & \\ p_x''' & \cdots & & \\ a & b & c & d \end{matrix} \right\Vert = 0, \qquad \left\Vert \begin{matrix} p_x & q_x & r_x & s_x & A \\ p_x' & \cdots & & & B \\ p_x'' & \cdots & & & C \\ p_x''' & \cdots & & & D \end{matrix} \right\Vert = 0.$$

Denoting these two kinds of curves by c_6 and k_6, it is found that any two curves of the same kind meet in four points, any two curves of different kinds in fourteen points. Any two curves of different kinds lie on a cubic surface.

The surface can be birationally transformed into itself by aid of the three sets of equations

$$\begin{aligned} &\lambda_1 p_x + \lambda_2 q_x + \lambda_3 r_x + \lambda_4 s_x = 0, && \alpha_1 p_y + \alpha_2 p_y' + \alpha_3 p_y'' + \alpha_4 p_y''' = 0, \\ &\lambda_1 p_x' + \cdots\cdots = 0, && \alpha_1 q_y + \cdots\cdots = 0, \\ &\lambda_1 p_x'' + \cdots\cdots = 0, && \alpha_1 r_y + \cdots\cdots = 0, \\ &\lambda_1 p_x''' + \cdots\cdots = 0; && \alpha_1 s_y + \cdots\cdots = 0; \end{aligned}$$

$$\lambda_1 P_1 + \lambda_2 Q_1 + \lambda_3 R_1 + \lambda_4 S_1 = 0,$$
$$\lambda_1 P_2 + \ldots\ldots\ldots\ldots\ldots\ldots = 0,$$
$$\lambda_1 P_3 + \ldots\ldots\ldots\ldots\ldots\ldots = 0,$$
$$\lambda_1 P_4 + \ldots\ldots\ldots\ldots\ldots\ldots = 0;$$

where $$P_i = \alpha_1 p_i + \alpha_2 p_i' + \alpha_3 p_i'' + \alpha_4 p_i'''.$$

If we regard the λ_i and the α_i as point-coordinates we pass, by aid of these equations, from a point x of Δ to a point λ of a surface Σ, thence to a point α of a surface Σ' and finally to a point y of Δ.

From the preceding equations we deduce that if x is any point of a curve c_6, the point x determines a trisecant of c_6 whose fourth intersection with Δ is the point y, which corresponds to x.

These trisecants, as x describes c_6, form a ruled surface of the eighth order, whose intersection with Δ is c_6 taken triply together with a curve of the fourteenth order, the locus of the points y on Δ corresponding to the points x of c_6.

When the determinant Δ is symmetrical, i.e. if

$$p' \equiv q, \quad p'' \equiv r, \quad p''' \equiv s, \text{ etc.},$$

the surfaces Σ and Σ' coincide; and the quantities P_i, Q_i, etc. are in this case the partial derivatives of a quantity which is quadratic in the α_i; if, changing the notation, we represent this quantity by S_i, the last set of equations take the form

$$\sum_{i=1}^{i=4} x_i \frac{\partial S_j}{\partial y_i} = 0 \quad \ldots\ldots\ldots\ldots\ldots\ldots\ldots\ldots(1),$$

on replacing λ and α by x and y respectively.

Thus the surface Σ is the Jacobian J, of four quadrics. The surface $\Delta = 0$, where Δ is a symmetrical determinant, is known as the *symmetroid*; if in the first set of preceding equations we replace x, λ, y and α by α, x, β and y respectively, and express that $q \equiv p'$, etc., these equations assume the form

$$\sum_{i=1}^{i=4} \alpha_i \frac{\partial S_i}{\partial x_j} = 0, \quad \sum_{i=1}^{i=4} \beta_i \frac{\partial S_i}{\partial y_j} = 0, \quad (j = 1, 2, 3, 4)\ldots\ldots(2).$$

The surface Δ, the locus of the points α, is obtained by eliminating the x_i, or the y_i, from these equations. The surface J is seen to be the locus of vertices of cones of the system

$$\sum_1^4 \alpha_i S_i = 0.$$

The equations (1) express that the polar planes of any point x of J, with regard to each of these quadrics $S_1 \dots S_4$, are concurrent in the point y of J; the points x, y are said to be *corresponding* points on J.

The surface Δ has ten nodes; the tangent cone of Δ whose vertex is any of its nodes breaks up into two cubic cones; a characteristic property of this surface.

The surface J has ten lines; every point of a line of J is associated with the *same* point α of Δ, by equations (2), which is a node of Δ.

The tangent plane of J at any point P is the polar plane of P', the point *corresponding* to P, for the cone of the system $\sum_1^4 \alpha_i S_i$ whose vertex is P.

When x describes a line of the Jacobian, its corresponding point y describes a twisted cubic; the point β on the symmetroid describes a curve of the ninth order having double points at each node of the symmetroid except the one which is connected with the locus of x.

As the point y describes the section of the Jacobian made by the plane $a_y = 0$, the corresponding locus of x is the sextic

$$\left\| \frac{\partial S_1}{\partial x_i} \dots \frac{\partial S_4}{\partial x_i},\ a_i \right\| = 0,$$

which has the ten lines of the Jacobian as trisecants. The locus of the associated points α on the symmetroid is a curve of the fourteenth order, passing three times through each node; that of the associated points β is a sextic curve which passes through the ten nodes. To a plane section through two nodes of the symmetroid there corresponds a quadri-quartic on the Jacobian.

If the quadrics $S_1 \dots S_4$ have a common point, the Jacobian has a node and an additional node arises on the symmetroid. Each additional common point of $S_1 \dots S_4$ will give rise to a node on both the Jacobian and the symmetroid. If there are six such common points, the Jacobian becomes the surface known as Weddle's, and the symmetroid becomes Kummer's surface.

Weddle's surface has thus the six points common to $S_1 \dots S_4$ as nodes, and contains twenty-five lines, viz. the fifteen lines joining the nodes and the intersections of the ten pairs of planes through the six points.

The line joining any two corresponding points P, P' of the surface meets the twisted cubic through the six nodes in two points L, M such that the four points P, P', L, M are harmonic. It follows that this cubic is an asymptotic line of the surface.

If θ^3, θ^2, θ, 1 be the coordinates of any point on this twisted cubic, then the coordinates of the preceding points P, P' are obtained as follows: let θ, ϕ denote the points L, M; the coordinates of P, P' are given by the equations

$$x_1 : x_2 : x_3 : x_4$$

$$= \frac{\theta^3}{\sqrt{f(\theta)}} \pm \frac{\phi^3}{\sqrt{f(\phi)}} : \frac{\theta^2}{\sqrt{f(\theta)}} \pm \frac{\phi^2}{\sqrt{f(\phi)}} : \frac{\theta}{\sqrt{f(\theta)}} \pm \frac{\phi}{\sqrt{f(\phi)}} : \frac{1}{\sqrt{f(\theta)}} \pm \frac{1}{\sqrt{f(\phi)}},$$

where $f(\alpha) = \prod_1^6 (\alpha - \theta_i)$, and $\theta_1 \ldots \theta_6$ are the values of θ relating to the six nodes.

Any two points θ, ϕ of the twisted cubic thus determine two points P, P' on the surface; any three points θ, ϕ, ψ determine three pairs PP', QQ', RR' of corresponding points which form the vertices of a complete quadrilateral; any four points θ, ϕ, ψ, χ determine twelve points which form three desmic tetrahedra, viz.

$$PP'SS', \quad QQ'TT', \quad RR'UU'.$$

If in the preceding expression of the points of the surface in terms of θ, ϕ we suppose θ to be constant, i.e. take all chords through a given point of the twisted cubic, the resulting locus of points of the surface is a quintic curve; these curves form a conjugate system on the surface. If the tangent to the twisted cubic at the point θ meets the surface again in the point T, then the locus of points of contact of the tangents from T to the surface is one of these curves.

The surface, being defined as the locus of vertices of cones which pass through six given points, is seen to have an equation of the form

$$\frac{p_{12}p_{34}}{p_{13}p_{42}} = \frac{q_{12}q_{34}}{q_{13}q_{42}},$$

provided that four nodes are taken as vertices of the tetrahedron of reference, and p_{ik}, q_{ik} are the coordinates of the lines joining any point of the surface to the two remaining nodes.

This equation expresses that the lines p, q meet the faces of the tetrahedron formed by the four nodes in two sets of four points which have the same anharmonic ratio.

It can be deduced that a form of the equation of the surface is

$$\begin{vmatrix} \frac{a_1 b_1}{x_1} & \frac{a_2 b_2}{x_2} & \frac{a_3 b_3}{x_3} & \frac{a_4 b_4}{x_4} \\ x_1 & x_2 & x_3 & x_4 \\ a_1 & a_2 & a_3 & a_4 \\ b_1 & b_2 & b_3 & b_4 \end{vmatrix} = 0.$$

Any point P of the surface determines a closed set of thirty-two points on the surface as follows: if P be joined to the six nodes $N_1 \dots N_6$, then calling the point of second intersection of PN_1 with the surface (N_1), etc., we thus obtain the six points $(N_1) \dots (N_6)$; secondly, by joining such a point (N_1) to the nodes, we obtain five points of second intersection (N_1N_2), etc.; there are fifteen such points; lastly, by joining the points (N_1N_2) to the nodes, we obtain the points $(N_1N_2N_3)$ which are only ten in number, since

$$(N_1N_2N_3) \equiv (N_4N_5N_6), \text{ etc.}$$

The surface may be shown to be a linear projection in four dimensions, and therefore projectively related to a Kummer surface. For the Weddle surface arises as the interpretation in three dimensions of the twofold of contact of the enveloping cone of a cubic variety in four dimensions, whose vertex is any point of the variety. Now, since the intersection of this cone with any arbitrary hyperplane is a Kummer surface, we are again led to a birational transformation between the Weddle and the Kummer surface.

The coordinates of any point of the surface can be expressed as being proportional to the ratios of the products of four double theta functions: viz. the substitutions

$$x_1 : x_2 : x_3 : x_4 = c_{01}\theta_{01}\theta_3\theta_{02}\theta_{04} : c_2\theta_2\theta_1\theta_3\theta_{04} : c_{03}\theta_{03}\theta_1\theta_{02}\theta_{04} : c_4\theta_4\theta_1\theta_3\theta_{02},$$

$$a_1 : a_2 : a_3 : a_4 = c_{01}c_0c_{23}c_{34} : c_2c_5c_{12}c_{23} : c_{03}c_0c_{12}c_{14} : c_4c_5c_{14}c_{34},$$

$$b_1 : b_2 : b_3 : b_4 = c_{01}c_5c_{12}c_{14} : c_2c_0c_{14}c_{34} : c_{03}c_5c_{23}c_{34} : c_4c_0c_{12}c_{23},$$

satisfy the equation of the surface.

We obtain two sets of quadri-quartics on the surface; the first set is given as the intersection of two cones passing through the same four nodes and having the other two nodes as respective vertices, viz. the cones

$$p_{12}p_{34} = \lambda p_{13}p_{42}, \quad q_{12}q_{34} = \lambda q_{13}q_{42};$$

the second set is given as the intersection of the quadrics

$$p_{12}p_{34} = \mu q_{12}q_{34}, \quad p_{13}p_{42} = \mu q_{13}q_{42};$$

each of these curves passes through four nodes; the equation of the last set, expressed in terms of double theta functions, is

$$\theta_{13} = \mu\theta_{24}.$$

The coordinates of the fifteen points (N_1N_2), etc. are obtained from those of any point for which the argument is u by the addition of one of the fifteen half-periods. The coordinates of the point (N_1) in which the join of P to A_1 meets the surface again are found to be

$$c_{01}\theta_2\theta_{03}\theta_4 : c_2\theta_2\theta_3\theta_{04} : c_{03}\theta_{03}\theta_{02}\theta_{04} : c_4\theta_4\theta_3\theta_{02}.$$

The fifteen other points (N_2), etc. and $(N_1N_2N_3)$, etc. are obtained by addition of one of the fifteen half-periods to the argument of u in these last expressions.

The equation of a plane section of the surface, referred to the three points in which the plane of section meets the twisted cubic through the six nodes, assumes a simple form. The tangents to the curve at the vertices of the triangle of reference meet in one point; an invariant of the curve is seen to vanish; the curve contains an infinite number of configurations of points, each configuration being formed by twenty-five points.

Bauer has investigated the surface whose equation is

$$\begin{vmatrix} x_1 - a_x/a_1 & x_2 & x_3 & x_4 \\ x_1 & x_2 - b_x/b_2 & x_3 & x_4 \\ x_1 & x_2 & x_3 - c_x/c_3 & x_4 \\ x_1 & x_2 & x_3 & x_4 - d_x/d_4 \end{vmatrix} = 0;$$

its origin is as follows: a point P is joined to the vertices of a tetrahedron (taken as that of reference) and the joining lines meet the faces of another tetrahedron (whose faces are $a_x = 0$, $b_x = 0$, $c_x = 0$, $d_x = 0$) in four points; if these latter points are coplanar, we obtain as locus of P the surface whose equation has just been given.

When the two tetrahedra are in perspective, the surface is the Hessian of the general cubic surface; it has ten nodes.

When the preceding connection mentioned in the beginning of the chapter, between the points x and y which gives rise to the surface Δ, reduces to a collineation, we obtain a surface, discussed by Schur, whose equation is

$$\alpha\beta\gamma\delta + \alpha'\beta'\gamma'\delta' = 0,$$

in which $\alpha \ldots \delta'$ are linear in the variables; and the collineation is such as to permute cyclically the planes $\alpha \ldots \delta$ and the planes $\alpha' \ldots \delta'$. This surface contains thirty-two lines.

If, in addition, the faces of both tetrahedra are subject to a collineation which leaves one face of each tetrahedron unaltered and permutes cyclically the other three, the surface contains fifty-two lines.

CHAPTER I

QUARTIC SURFACES WITH ISOLATED SINGULAR POINTS

1. The singular points possessed by a quartic surface may consist either of a certain number of isolated nodes or may form double curves.

In the present chapter we discuss the quartic surfaces which have an assigned number of nodes, beginning with those which have four nodes, and give a definite method of classification for all the cases in which the number of nodes exceeds seven.

The number of isolated nodes of a quartic surface cannot exceed sixteen; for the class of a surface of order n which has δ double points is $n(n-1)^2 - 2\delta$, since this is the number of points of intersection of the surface and its first polars for two points A and B, diminished by the number 2δ of these intersections arising from each double point (a simple point on the polars of both A and B). Hence if n is four, δ cannot exceed sixteen.

2. Quartic surfaces with four to seven nodes.

Since the equation of the general quartic surface contains thirty-four constants, the surface with four given nodes should contain $34 - 16 = 18$ constants; if then $A = 0$, $B = 0$, $C = 0$, $D = 0$, $E = 0$, $F = 0$ are six linearly independent quadrics through the four nodes, the equation

$$(a \between A, B, C, D, E, F)^2 = 0,$$

containing apparently twenty constants, is a quartic surface having the given nodes.

The number of constants is really eighteen, since there are two quadratic relations between the six quadrics, as may be seen by taking the four given points as vertices of the tetrahedron of reference, in which case the quadrics may be taken to be

$$x_1x_2, \quad x_1x_3, \quad x_1x_4, \quad x_3x_4, \quad x_2x_4, \quad x_2x_3,$$

between which there exist the identities

$$x_1x_2 \,.\, x_3x_4 = x_1x_3 \,.\, x_2x_4 = x_1x_4 \,.\, x_2x_3.$$

For five nodes, taking $A \dots E$ as quadrics passing through the given nodes, the equation

$$(a \between A, B, C, D, E)^2 = 0,$$

containing fourteen constants, represents the general quartic having these given nodes.

The general quartic with six nodes is represented by the equation

$$(a \between A, B, C, D)^2 + \rho J = 0,$$

where A, B, C, D are quadrics through the six nodes, and J is the Jacobian of the four quadrics. For this equation contains ten constants and J has the given points $D_1 \dots D_6$ as nodes, moreover J cannot be expressed as a quadratic function of A, B, C, D.

The following properties of J may be used to establish these results. The surface $J=0$ is the locus of vertices of cones of the system

$$A + \lambda B + \mu C + \nu D = 0;$$

now each point of the line joining any two double points, e.g. $D_1 D_2$, is the vertex of such a cone, hence J contains the join of any two double points; also since $D_1 D_2 \dots D_1 D_6$ lie on J it follows that D_1 is a node of J; similarly for $D_2 \dots D_6$. Again there are ten pairs of planes passing through the points $D_1 \dots D_6$, and each point of the line of intersection of such a pair of planes satisfies the condition of being the vertex of a cone of the system. Hence such a line lies upon J, which thus contains $15+10=25$ lines. Again, since any quadric of the system is linearly expressible in terms of any four members of the system, it is so expressible in terms of any four of the previous pairs of planes; hence if J were expressible as a quadratic function of A, B, C and D, we should necessarily have a relation of the form

$$J \equiv (a \between \alpha\alpha', \beta\beta', \gamma\gamma', \delta\delta')^2,$$

in which we may take the planes α, β, γ to contain the line $D_1 D_2$, while δ, δ' do not contain it, e.g.,

$$\alpha \equiv (D_1, D_2, D_3), \quad \alpha' \equiv (D_4, D_5, D_6), \quad \text{etc.,}$$

while $$\delta \equiv (D_1, D_3, D_5), \quad \delta' \equiv (D_2, D_4, D_6).$$

Hence since J contains $D_1 D_2$ such a relation is impossible.

The general quartic with seven nodes is represented by the equation

$$(a \between A, B, C)^2 + \rho \Sigma = 0,$$

where A, B, C are quadrics through the given nodes and Σ is any quartic surface having the seven nodes*.

* This quartic surface may also be expressed in terms of the quartic surfaces which have one of the given points as a triple point and the other six as double points; if $T_1 \dots T_7$ are these surfaces, the required general quartic surface is $\Sigma a_i T_i = 0$, $i = 1, \dots 7$.

3. Quartic surfaces with more than seven nodes.

The equation of a quartic surface having a node at the point $x = y = z = 0$ will be of the form

$$u_2 w^2 + 2u_3 w + u_4 = 0,$$

where $u_2 = 0$, $u_3 = 0$, $u_4 = 0$ are cones whose vertex is the node.

The equation of the tangent cone drawn to the surface from this node is

$$u_2 u_4 - u_3^2 = 0.$$

The section of this cone by any plane, e.g. the plane $w = 0$, is a sextic curve with a "contact-conic," i.e. a conic which touches it wherever it meets it.

If the surface has any other node, the tangent cone will have a double line passing through this new node and giving rise to a node on this sextic; we obtain the different varieties of quartic surfaces possessing nodes by consideration of all special cases of sextic curves with a contact-conic*.

It is to be noted that the existence of a contact-conic $u_2 = 0$ of a sextic implies also a contact-quartic $u_4 = 0$; if a sextic has another contact-conic $v_2 = 0$, and hence another contact-quartic $v_4 = 0$, an identity exists of the form

$$u_2 u_4 - u_3^2 \equiv v_2 v_4 - v_3^2.$$

Now by multiplying the equation of the surface by u_2 we derive

$$(u_2 w + u_3)^2 + u_2 u_4 - u_3^2 = 0,$$

hence in the present case

$$(u_2 w + u_3)^2 + v_2 v_4 - v_3^2 = 0 \quad \text{.................}(1).$$

Denoting by c_8 the intersection of the quartic surface with the cone $v_2 = 0$, it is clear that $v_2 = 0$ meets the surface (1) in the curve c_8 and in the four lines $u_2 = v_2 = 0$; but v_2 meets (1) where it meets the two nodal cubic surfaces

$$u_2 w + u_3 - v_3 = 0,$$

$$u_2 w + u_3 + v_3 = 0,$$

hence in general c_8 must break up into two quartic curves, either of which is the partial intersection of v_2 with a cubic surface which contains also two generators of v_2. These curves are therefore quadri-quartics†. Hence the surface contains an infinite number

* This method is due to Rohn, see *Die Flächen vierter Ordnung hinsichtlich ihrer Knotenpunkte und ihrer Gestaltung*, Leipzig, 1886.

† We denote by *quadri-quartic* the type of twisted quartic through which an infinite number of quadrics pass.

of quadri-quartic curves which are projected from the node into quartic curves which touch the sextic $u_2u_4 - u_3^2 = 0$ at each point of intersection*.

Hence if the curve $u_2u_4 - u_3^2 = 0$ has more than one contact-conic it has an infinite number of contact-conics.

4. Nodal sextics†.

For the purpose of classification of nodal quartic surfaces we discuss various properties of sextic curves with a contact-conic. In the first place it may be seen that sextic curves with *six nodes* lying on a conic c_2 can have their equation expressed as above. For if $c_3 = 0$ is any cubic through the six points, any other cubic through them is of the form $c_3 + c_2L = 0$; and any sextic through the complete intersection of c_2 and c_3 being

$$c_2c_4 + c_3c_3' = 0,$$

if the six points are nodes on this sextic c_4 and c_3' must be of the form $c_3M + c_2N$, $c_3 + c_2R$ respectively.

Hence the required sextic takes the form

$$c_3^2 + c_2c_3A + c_2^2B = 0,$$

i.e. the form $$K_3^2 - c_2^2V = 0,$$

and hence has a contact-conic.

The corresponding quartic surface is $w^2V + 2wK_3 + c_2^2 = 0$; this has the plane $w = 0$ as a singular tangent plane or *trope*, which touches the surface along a conic.

Sextics with seven nodes.

There are two different kinds of seven-nodal sextics, viz. that for which it is possible to find a pair of points P, P' on the curve, such that through the seven nodes $D_1 \ldots D_7$ and P, P' there pass an infinite number of cubics, and the one for which it is not possible; considering the former kind, then if one such pair of points exists there is an infinite number of such pairs; for taking c_3 and c_3' as two such cubics, then c_6, the given sextic, since it passes through the complete intersection of c_3 and c_3', has an equation of the form

$$c_3\Gamma_3 + c_3'\Gamma_3' = 0.$$

* For such a point of intersection P is the projection of an actual intersection Q of the quadri-quartic and the curve of contact of the tangent cone, and the tangents to these curves at Q lie in the tangent plane of the surface.

† See Rohn, *l.c.*

Now c_3 meets c_6 only in $D_1 \dots D_7$, P, P' and two further points Q, Q', hence Γ_3' passes through $D_1 \dots D_7$ and also through Q and Q'; so that two and therefore an infinite number of cubics pass through $D_1 \dots D_7$, Q and Q'. By varying the cubic through the nine points $D_1 \dots D_7$, P, P' we form an involution of points Q, Q' on c_6. If Q coincides with P, Q' will coincide with P'; therefore every cubic through the seven nodes which touches c_6 once will touch it twice.

Since Γ_3 is seen to pass through $D_1 \dots D_7$ and since only three linearly independent cubics pass through seven points, there is a linear connection between c_3, c_3', Γ_3 and Γ_3', hence the sextic which has the property considered is represented by an equation of the form

$$(a \between \phi, \psi, \chi)^2 = 0,$$

where ϕ, ψ, χ are any three cubics through the given nodes.

This class of sextic always has a contact-conic; for if the sextic is $c_6 \equiv c_3^2 - c_3' c_3''$, let the chord joining the intersections P_1, P_2 of c_3' and c_3'', apart from the nodes, be $f = 0$, and $f' = 0$, $f'' = 0$ similar chords for c_3'', c_3 and c_3, c_3'; then fc_3, $f'c_3'$, $f''c_3''$ all pass through the thirteen points $D_1 \dots D_7$, P_1, $P_2, \dots P_2''$, and hence through three other fixed points*. Hence we have a linear relation of the form

$$Afc_3 + Bf'c_3' + Cf''c_3'' = 0,$$

where A, B, C are definite constants.

Now if $\quad c_4 \equiv -\tfrac{1}{2}Afc_3 - Bf'c_3' \equiv \tfrac{1}{2}Afc_3 + Cf''c_3''$,

we have $\quad c_4^2 - \tfrac{1}{4}A^2f^2c_3^2 + BCf'f''c_3'c_3'' \equiv 0;$

that is $\quad 4c_4^2 - A^2f^2c_6 + c_3'c_3''(4BCf'f'' - A^2f^2) \equiv 0.$

Hence the conic $4BCf'f'' - A^2f^2 = 0$ touches c_6, viz. at six of its intersections with c_4, the other two being the points (f, f'), (f, f'').

This conic is touched by f' and f'', hence *the tangents of the contact-conic are the chords of contact of c_6 and its bitangent cubics.*

We observe that in this case there is a doubly infinite number of quartic curves c_4 which pass through the seven nodes and the six points of contact of c_6 and its contact-conic.

* Since all quartics through thirteen points which do not all lie on a curve of lower degree pass through three other fixed points and hence belong to a pencil.

Sextics with eight or with nine nodes.

If $f=0$ is *any* sextic with eight nodes $D_1 \dots D_8$ and $\phi=0$, $\psi=0$ any two cubics through them, the general sextic with the eight given nodes is

$$f+\lambda\phi^2+\mu\phi\psi+\nu\psi^2=0.$$

If this curve has a ninth node it either degenerates into two cubics through the nine points (which are then the complete intersection of two cubics) or the ninth node lies on the curve $J(f, \phi, \psi)=0$. This is of the ninth degree and will be denoted by c_9; it has each of the eight nodes as a triple point*.

The curve $f=0$ and the eight nodes completely determine c_9; if we take any point P of intersection of f and c_9, and suppose ϕ to pass through P, then any sextic with the eight given nodes is of the form $f+\rho\phi\phi'=0$ where ϕ' does not pass through P.

Since P lies on c_9 it follows from the equation of that curve that the linear polars of P for f, ϕ and ϕ' concur; but the first two are the tangents at P to f and ϕ, and the third cannot pass through P, hence f and ϕ touch at P, and ϕ touches every sextic with the eight given nodes which pass through P. Now f and c_9 meet in $9\times6-8\times6=6$ points apart from the nodes, hence every sextic *with eight nodes is touched by six cubics through these nodes.*

If $f=0$ is any sextic with nine nodes and $\phi=0$ the cubic through them, $f+\rho\phi^2=0$ is the equation of the general sextic with the given nine nodes. If there is a tenth node it will be included among the points determined by the equations

$$\left\| \begin{matrix} f_1 & f_2 & f_3 \\ \phi_1 & \phi_2 & \phi_3 \end{matrix} \right\| = 0.$$

The number of solutions given by these equations is thirty-nine, but each of the given nine nodes occurs as a triple solution. Hence the pencil of sextics $f+\rho\phi^2=0$ contains twelve curves which have a tenth node (see Art. 9).

The foregoing result as to contact-cubics is modified as follows: through any eight nodes of a sextic with nine nodes there pass four tangent cubics; through any eight nodes of a sextic with ten nodes there pass two tangent cubics.

5. Sextics with ten nodes.

The following result for ten-nodal sextics is important for our purpose: *every plane sextic with ten nodes and a contact-conic is the projection of a twisted sextic on a quadric*: for choosing any centre

* As may be seen by taking any one of them as $x=0$, $y=0$, $z=0$.

of projection O and any quadric whose section by the polar plane of O for the quadric projects into the given contact-conic, the sextic cone whose base is the given sextic meets the quadric in a curve c_{12} which has twenty-six actual double points, since each node of the plane sextic gives rise to two nodes on c_{12}, and each point of contact of the contact-conic and the sextic is the projection of a point at which two branches of c_{12} touch each other. Moreover c_{12} has thirty apparent double points*, hence the projection of c_{12} from any point has $30+26=56$ nodes, and this is one more than can be possessed by a curve of order 12 which does not break up into simpler curves. Hence c_{12} must break up into two sextic curves.

There are *three* varieties of twisted sextics on a quadric: (1) its intersection with a cubic surface, (2) its partial intersection with a quartic surface which also contains two generators of the quadric of the same species, (3) its partial intersection with a quintic surface which also contains four generators of the quadric of the same species.

The following result, which may be easily proved†, is of frequent application: through every point P of space there pass $n(n-1)$ double secants of the complete curve of intersection of a quadric with any surface of order n; these double secants form the intersection of a cone of order n with a cone of order $n-1$, the former cone passes through the $2n$ intersections of the polar plane of P and this curve.

Let us now consider the plane ten-nodal sextic which is the projection of the first of these three varieties. This has six apparent double points and, since its plane projection has ten

* Salmon, *Geom. of three dimensions* (fifth ed. 1912), vol. I. p. 356.

† If $V=0$ is the surface and $U=0$ the quadric, it is easy to see that the section of the curve of intersection by the polar plane of P for U is given by the equations

$$\Delta U=0, \quad \left(V-\frac{U}{U'}\frac{\Delta^2 V}{\lfloor 2}+\ldots\right)^2+\frac{U}{U'}(\Delta V-\ldots)^2=0,$$

where

$$\Delta U=\sum_1^4 x_i'\frac{\partial}{\partial x_i}U,$$

and x_i' are the coordinates of P. Relatively to its plane the equation of this curve is of the form

$$v_n{}^2+c_2 v^2{}_{n-1}=0;$$

this curve contains $n(n-1)$ nodes which arise solely from apparent double points of the curve $U=0$, $V=0$; also $v_n=0$ is seen to pass through the common intersections of $V=0$, $U=0$, $\Delta U=0$.

nodes, it must have four actual double points; by the last result six of the nodes of the plane sextic lie on a conic; it is therefore represented by an equation of the form

$$K_3^2 - c_2^2 V = 0. \qquad \text{(Art. 4.)}$$

The second species of twisted sextic lies on a quadric and a quartic surface, their intersection being completed by two generators of the quadric. This curve has seven apparent double points*; and therefore, to complete the number of nodes of the plane quartic, must have three actual double points. Each generator of the given species meets the curve four times. There is an infinite number of quartic surfaces passing through the sextic and any two generators of the quadric. For any quartic surface through five points of each generator and any seventeen points of the sextic will meet the sextic in $8 + 17 = 25$ points, and therefore contain it altogether: it will also contain the two generators. Let us denote the twisted sextic by c_6, its plane projection by c_6', and take any generator p and its consecutive generator as the pair of generators just mentioned; then the *cubic* cone which contains the seven double secants of c_6 will touch c_6' twice†; hence, varying p, we obtain *an infinite number of cubics through seven nodes of* c_6' *and bitangent to it.*

In the third type of twisted sextic c_6 is the partial intersection of a quadric and a quintic, the residual intersection being formed by four generators of the quadric of the same species. Each generator of this species meets the sextic five times. It may be shown as before that there is an infinite number of quintic surfaces passing through the given sextic and any four generators of the given species. The curve c_6 has ten apparent double points.

We may select the four generators as follows: let p and p' be those generators which are projected from the centre of projection O into the tangents of the contact-conic of c_6' drawn from some node D of c_6'; we then take as our four generators p, p' and the generators consecutive to them. The line OD thus meets c_6 twice, and serves as join of apparent intersections for c_6, p and for c_6, p'. The compound curve of intersection of order 10 has twenty apparent double points, of which nine are projected into D, viz. one point arising from c_6, two from $(c_6, p)(c_6, p+dp)$, two from $(c_6, p')(c_6, p'+dp')$ and four from p and p'.

Hence the two cones of orders 4 and 5 through the double secants must

* After deduction of five apparent double points arising from the two lines.

† Since p gives rise to two apparent double points of the compound curve.

each have a common triple edge; we therefore obtain the following results: if D is any one of the ten nodes there exists a quartic curve which has a triple point in D and passes through the nine other nodes and the points of contact of the tangents drawn from D to the contact-conic; also there exists a quintic curve which has a triple point in D, passes through the nine other nodes and touches the contact-conic where it is touched by its tangents drawn from D. This holds for each node.

6. Quartic surfaces with eight nodes.

Returning to the sextic curve $u_2u_4 - u_3^2 = 0$, derived from the surface $u_2w^2 + 2u_3w + u_4 = 0$, any quadric through the node is

$$2t_1 w + t_2 = 0;$$

if the quartic surface has any other node which also lies upon this quadric, since this node also lies on the surface

$$u_2 w + u_3 = 0,$$

it is clear that the curve $u_2t_2 - 2t_1u_3 = 0$ will pass through the resulting node on $u_2u_4 - u_3^2 = 0$ or c_6.

This quartic curve passes through the points of contact of c_6 with its contact-conic u_2, and also through the nodes of c_6 which result from nodes on the quartic surface. If therefore the surface has eight nodes we have seven nodes on c_6: to each quartic through these seven nodes and the points of contact $B_1 \ldots B_6$ of c_6 and u_2, there corresponds one quadric through the eight nodes, and vice-versâ.

Now it was stated (Art. 4) that plane sextics with seven nodes form two classes; in the more general case there is a singly infinite number of quartic curves through the nodes and $B_1 \ldots B_6$, and we obtain corresponding to this case a singly infinite number of quadrics through the eight nodes. For the more special case where there is a doubly infinite number of quartic curves through the thirteen points we have a doubly infinite number of quadrics through the eight nodes, which therefore form *eight associated points.* Such a surface is represented by an equation of the form

$$(a \between A, B, C)^2 = 0.$$

It follows that any quadric through the eight nodes meets the quartic surface in two quadri-quartic curves which are projected from any node into two of the ∞^2 cubics which pass through the seven nodes of c_6.

These two classes of quartic surfaces will be termed *asyzygetic* and *syzygetic* respectively*.

The equation of the general seven-nodal surface being $F \equiv (a \between A, B, C)^2 + \rho\Sigma = 0$ (Art. 2), where A, B, C are quadrics through the seven nodes, if there is an eighth node we obtain, to determine it, the equations

$$A_i \frac{\partial F}{\partial A} + B_i \frac{\partial F}{\partial B} + C_i \frac{\partial F}{\partial C} + \rho\Sigma_i = 0, \quad (i = 1, \ldots 4);$$

hence the eighth node lies on the surface

$$J(A, B, C, \Sigma) \equiv \begin{vmatrix} A_1 & B_1 & C_1 & \Sigma_1 \\ A_2 & B_2 & C_2 & \Sigma_2 \\ A_3 & B_3 & C_3 & \Sigma_3 \\ A_4 & B_4 & C_4 & \Sigma_4 \end{vmatrix} = 0.$$

The eighth node may therefore not be taken arbitrarily, as in the case of the first seven nodes.

If A, B are two quadrics through the eight nodes and T any eight-nodal asyzygetic surface, the general asyzygetic surface is represented by the equation

$$\alpha A^2 + \beta B^2 + 2\gamma AB + 2\rho T = 0.$$

The surface J is called the *dianodal surface*†, and is the locus of a point whose polar planes for A, B, C and Σ are concurrent, and therefore also concurrent for every quartic surface with the given seven nodes; thus if P is any point of the dianodal surface, all the quartics through P have a common tangent line thereat, which touches the quadri-quartic through P and the seven nodes, as is seen by taking as the quartic a doubled quadric through P and the seven nodes.

The dianodal surface.

The dianodal surface contains the line joining any two nodes D_1, D_2; for if P be any point on this line then, since we may take the surfaces A, B, Σ which appear in the equation of the seven-nodal quartic to pass through P, they will necessarily contain the line D_1D_2, hence the tangent planes at P to A, B, Σ all pass through D_1D_2 and therefore the point P satisfies the equation of

* The general syzygetic surface is the envelope of the quadrics $\lambda^2 D + \lambda E + F = 0$, where D, E, F are quadrics through the eight nodes.

† Cayley.

the dianodal surface. This surface thus contains the twenty-one lines which join any two of the nodes $D_1 \dots D_7$.

Again taking Σ, A and B to pass through any given seventh point of the twisted cubic determined by $D_1 \dots D_6$, this cubic lies entirely in Σ as meeting it in thirteen points, and also on A and B as meeting them in seven points, hence the tangent planes at P to Σ, A and B will meet in the tangent line at P to the twisted cubic: hence, as before, the point P lies on the dianodal surface. This surface thus contains the seven twisted cubics which pass through any six of the points $D_1 \dots D_7$.

The dianodal surface contains thirty-five plane cubics lying on the planes which contain three of the given nodes; for let L be the plane of three nodes and S the cubic surface which passes through these three nodes and has the four other nodes as double points; if we then write $L \,.\, S$ for Σ in the equation of the dianodal surface it becomes

$$J(A, B, C, L \,.\, S) \equiv LJ(A, B, C, S) + SJ(A, B, C, L) = 0,$$

which clearly contains the cubic $L = 0$, $S = 0$. This shows that the lines D_1D_2, etc. are simple lines of the dianodal surface.

The twisted sextic which is the locus of the vertices of the cones which pass through the seven given nodes, lies on the surface; for this sextic is obtained by elimination of λ, μ from the equations

$$A_i + \lambda B_i + \mu C_i = 0, \qquad (i = 1, \dots 4),$$

which clearly lies upon $J(A, B, C, \Sigma) = 0$.

Each of the seven nodes is a triple point of J, for the lines $D_1D_2, \dots D_1D_7$ do not lie on the same quadric cone.

7. Quartic surfaces with nine nodes.

From the two varieties of surfaces with eight nodes we derive two with nine nodes. Considering first syzygetic surfaces, viz.

$$\alpha A^2 + \dots + 2\mu AB = 0,$$

if this surface has a ninth node it must lie upon the twisted sextic

$$\left\| \begin{matrix} A_1 \dots A_4 \\ B_1 \dots B_4 \\ C_1 \dots C_4 \end{matrix} \right\| = 0.$$

This curve is the locus of the vertices of the cones of the system

$A + \lambda B + \mu C$; hence, if a ninth node exist, there is a quadric cone K whose vertex is D_9 which passes through the points $D_1 \dots D_8$.

Taking D_9 as the point from which the surface is projected by a tangent cone (giving rise to the curve $u_2u_4 - u_3^2 = 0$), this latter curve must have eight nodes lying on a conic, and must therefore break up into this conic and a quartic curve. Therefore K, the quadric cone whose vertex is D_9, forms part of the tangent cone from D_9, and touches the quartic surface along a twisted quartic. The equation of the surface is therefore of the form

$$A^2 + \rho KB = 0,$$

where $A = 0$ is a quadric through the nine nodes and $B = 0$ a quadric through the eight associated points $D_1 \dots D_8$. There is a triply infinite number of nine-nodal syzygetic quartic surfaces.

Considering next asyzygetic nine-nodal surfaces, from the equation of the general eight-nodal surface it is seen that a ninth node must lie on the curve

$$\begin{Vmatrix} A_1 \dots A_4 \\ B_1 \dots B_4 \\ T_1 \dots T_4 \end{Vmatrix} = 0,$$

which is of the eighteenth order, the *dianodal curve*; the ninth node being taken arbitrarily on this curve, there is a singly infinite number of surfaces with the nine given nodes represented by the equation

$$A^2 + \rho P = 0,$$

where A is the quadric through the nine nodes and P any quartic surface with these nodes.

The dianodal curve.

The dianodal curve lies on each of the eight dianodal surfaces obtained from the eight given nodes; moreover the dianodal surfaces corresponding to $D_1 \dots D_6 D_7$ and $D_1 \dots D_6 D_8$ intersect in the fifteen lines joining any two of the points $D_1 \dots D_6$, in the dianodal curve, and in the twisted cubic through $D_1 \dots D_6$.

Through D_1, as being a triple point on each, there pass nine branches of the curve of intersection of the two dianodal surfaces, but of these, six branches arise from the lines $D_1D_2 \dots D_1D_6$ and the tangent at D_1 to the cubic $D_1 \dots D_6$; the remaining three branches arise from the dianodal curve which has therefore a triple point in each of the eight nodes $D_1 \dots D_8$.

Moreover since six of the intersections of the tangent cubic cones at D_1 to the two dianodal surfaces lie on the quadric cone of vertex D_1 and passing through $D_2 \dots D_6$, it follows that the remaining three intersections must lie in a plane, hence the tangents to the three branches of the dianodal curve at D_1 are coplanar.

The dianodal curve is seen from its equation to be the locus of a point whose polar planes for A, B and T are *coaxal*. In its equation we may take A to be the quadric through the eight nodes and any assigned point P, then B will not pass through P, and if T is a quartic of the system which passes through P, then if P is on the dianodal curve, since the polar plane of B for P cannot pass through P, it follows that A and T have the same tangent plane at P.

We may also note the following results: (1) the dianodal curve meets each of the lines D_1D_2 twice, apart from D_1 and D_2, (2) it meets each of the seven twisted cubics $D_1 \dots D_6$, etc. twice, apart from the nodes. For we may take the quadric A and the quartic T as passing through any point P of the line D_1D_2 which will then lie on each of them, hence we have at each point of D_1D_2 a (1, 1) correspondence of tangent planes which involves two coincidences, say at the points Q and Q', thus both Q and Q' satisfy the equation of the dianodal curve. Next take A and T as passing through some assigned point P of the cubic through $D_1 \dots D_6$; this cubic will then lie on each of these surfaces, so that they will also meet in a residual quintic curve which passes through the points $D_1 \dots D_6$. Now the number of points of apparent intersection of these curves is seen to be seven* and hence their actual intersections are eight in number, and deducting the six points $D_1 \dots D_6$ we obtain *two* as the number of their intersections apart from the nodes; at each of these points A and T touch, and hence each point lies on the dianodal curve.

8. Quartic surfaces with ten nodes.

We have, as before, two classes of irreducible sextics with nine nodes, viz. according as the points of the curve are or are not *conjugate* in pairs with regard to any seven of the nine nodes. We have also the sextic arising from two cubics or two lines and a quartic. We consider in the first place these last two cases.

* Salmon, *Geom. of three dimensions* (fifth ed.), vol. I. p. 358.

It may be shown that, if the tangent cone from one node of a ten-nodal quartic surface breaks up into two cubic cones, this will also occur for each node. For let the tangent cone from D_1 break up into the cubic cones V and V', touching the surface along the curves c_6 and c_6' respectively, then D_1 is a triple point on both c_6 and c_6'*, and $D_2 \ldots D_{10}$ are ordinary points on c_6 and c_6'. Now the cubic surface which has D_1 and D_2 for nodes and which passes through $D_3 \ldots D_{10}$ and also through any other three points on c_6, will meet c_6 in $6 + 2 + 8 + 3 = 19$ points and therefore contain c_6; it therefore meets the quartic surface in another curve k_6 which has D_2 as triple point and $D_1 D_3 \ldots D_{10}$ as ordinary points. Hence k_6 is projected from D_2 by a *cubic* cone which passes through D_1, D_2, $D_3 \ldots D_{10}$.

In the same manner, by aid of c_6', we obtain another sextic curve k_6' which projects from D_2 by a cubic cone. Hence the lines D_2D_1, $D_2D_3 \ldots D_2D_{10}$ form the complete intersection of two cubic cones, so that the sextic tangent cone to the quartic surface from D_2 has as double edges the complete intersection of two cubic cones: it must therefore break up into two cubic cones. Applying the same reasoning to each node it is seen that the tangent cone from each of them must break up into two cubic cones. This surface is called the *symmetroid*†.

In the next place, when the sextic splits up into two lines and a quartic curve, we see that through the node $x = y = z = 0$ there pass two planes, each touching the surface along a conic; each is a trope. The equation of the surface is of the form

$$A^2 + \rho xyB = 0,$$

where A and B are any two quadrics.

* Since any plane through D_1 meets c_6 in three points apart from D_1 and so for c_6'.

† See chap. IX. It is seen from the foregoing that any cubic cone whose vertex is a node and which passes through the nine other nodes, meets the surface in two sextic curves having the vertex as triple point and passing through the nine nodes.

We thus obtain ten sets of sextic curves on the surface.

Since the equation of the surface may be written

$$0 = M^2 + VV' = u_2 \,.\, F,$$

where $F = w^2u_2 + 2wu_3 + u_4,\quad M = wu_2 + u_3,\quad VV' = u_2u_4 - u_3^2,$

it follows that the cubic surfaces

$$\rho^2V + 2\rho M - V' = 0$$

touch F along the sextics $F = 0$, $\rho^2V + V' = 0$.

The nodes lie on two conics: the tangent cone from each of the eight associated nodes breaks up into a plane and a quintic curve with four double points.

We now pass to irreducible sextics, first those whose points are conjugate in pairs giving syzygetic surfaces with ten nodes. Such surfaces are represented by an equation of the form

$$A^2 + \rho K_1 K_2 = 0,$$

where K_1 and K_2 are cones whose vertices lie on the quadric A. Next if $P = 0$ is any asyzygetic surface with nine nodes, then among the surfaces

$$A^2 + \rho P = 0,$$

there are thirteen which have a tenth node; for such a node is an intersection of the dianodal surface of $D_1 \dots D_6 D_9$ and the dianodal curve of $D_1 \dots D_8$; there are $6 \times 18 = 108$ such intersections, but of these $D_1 \dots D_6$ being triple points on both the surface and the curve count as $9 \times 6 = 54$ intersections, and the points D_8, D_7, D_9 each count as three, also the two intersections of the fifteen lines $D_1 D_2$, etc. with the dianodal curve give thirty points, and its two intersections with the twisted cubic $D_1 \dots D_6$ give two more points which are not solutions; this leaves

$$108 - 54 - 9 - 30 - 2 = 13^* \text{ solutions.}$$

* Of these thirteen solutions *one* gives a symmetroid; for if P and A have the equations

$$w^2 u_2 + 2w u_3 + u_4 = 0, \quad w t_1 + t_2 = 0,$$

where D_1 is the point $x = y = z = 0$, we may write the equation $A^2 + \rho P = 0$ in the form

$$w^2 (t_1^2 + 2\rho u_2) + 2w (t_1 t_2 + 2\rho u_3) + t_2^2 + 2\rho u_4 = 0;$$

the sextic curve is therefore

$$2\rho (u_2 u_4 - u_3^2) + c_6 = 0,$$

where $$c_6 \equiv t_1^2 u_4 + t_2^2 u_2 - 2t_1 t_2 u_3$$

is the projection of the curve of intersection of A and P. All these curves have as double points the projections $D_2' \dots D_9'$ of $D_2 \dots D_9$, and c_6 has also as double points those in which the generators of A through D meet the plane of projection. All these curves touch c_6 twice.

Now all sextics having as nodes $D_2' \dots D_9'$ and which touch c_6 twice must have an equation either of the form

$$c_6 + \sigma \phi_3^2 = 0,$$

where ϕ_3 is a cubic through $D_2 \dots D_9$, or of the form

$$c_6 + \sigma \phi_3 \chi_3 = 0,$$

where ϕ_3, χ_3 are two cubics through the nodes $D_2' \dots D_9'$ of c_6 which touch it.

But the first form is excluded, since no doubled cubic can occur in the pencil of sextics; and the second form shows that as one curve of the pencil we have two cubics, i.e. for one of the surfaces $A^2 + \rho P = 0$ the tangent cone from D_1 breaks up into two cubic cones, and we have a symmetroid.

9. Quartic surfaces with eleven nodes.

The three varieties of plane sextics with ten nodes (Art. 5) lead to three types of quartic surface with eleven nodes. The equation of the first variety was seen to be of the form

$$u_3^2 - u_2^2 K = 0;$$

this sextic arises from the quartic surface

$$Kw^2 + 2u_3 w + u_2^2 = 0.$$

The six nodes which lie on a conic are given by the equations $w = 0$, $u_2 = 0$, $u_3 = 0$; the plane $w = 0$ is a *trope*.

The tangent cone drawn to the surface from any one of these nodes breaks up into the plane $w = 0$ and a quintic cone, the tangent cones from the remaining five nodes are irreducible.

If $P = 0$ be a quartic surface having the six coplanar points as nodes and also five other nodes, and A a quadric passing through four of these last five nodes and also the conic containing the six nodes, then

$$P + \rho A^2 = 0$$

is a pencil of quartic surfaces having ten nodes: the equations

$$\left\| \begin{matrix} P_i \\ A_i \end{matrix} \right\| = 0$$

give forty solutions, but the given ten nodes count triply among them, leaving ten surfaces of the pencil having eleven nodes and of the type just mentioned. This surface may be called XI_c.

The second kind of plane sextic with ten nodes has an infinite number of bitangent cubics through seven of its nodes (Art. 5); the quartic surface to which it corresponds must therefore be syzygetic; the equation of the ten-nodal syzygetic surface being $A^2 + \rho K_1 K_2 = 0$ (Art. 8) it may be shown that in this pencil there are twelve surfaces which have an eleventh node. It is easy to see that the equation of such a surface has the form

$$\sqrt{K_1} + \sqrt{K_2} + \sqrt{K_3} = 0,$$

where $K_1 = 0$, $K_2 = 0$, $K_3 = 0$ are cones, and such that the vertex of K_1 lies upon $K_2 - K_3 = 0$, etc. This surface is called XI_b.

There remain two cases in which the sextic curve breaks up into simpler curves: either into two lines and a nodal quartic or into two cubic curves.

In the first case the equation of the surface is

$$A^2 + \rho Kxy = 0,$$

where A is a quadric and K a cone whose vertex lies on A. This is XI_d.

In the second case we have a symmetroid, hence

$$(u_2 w + u_3)^2 + v_3 v_3' \equiv u_2 (u_2 w^2 + 2u_3 w + u_4).$$

Here either v_3 or v_3' has a nodal line, arising from an eleventh node on the surface. The tangent cone from this eleventh node to the surface gives a plane sextic of the third variety. This case is XI_a.

10. Quartic surfaces with twelve nodes.

A surface with twelve nodes gives rise to a sextic curve with eleven nodes: this sextic must therefore break up into simpler curves. The cases which provide eleven nodes are the following:

(1) a quintic with six nodes, and a straight line,

(2) a quartic with two nodes, and two straight lines,

(3) two nodal cubics,

(4) a cubic, a conic and a straight line,

(5) a quartic with three nodes, and a conic.

It may be shown that a plane quintic with six nodes and a contact-conic may be regarded as the projection of a twisted quintic on a quadric. The proof is exactly similar to that for the plane sextic with ten nodes. By addition of a generator it is easy to see that we obtain a special case of the second class of twisted sextics on a quadric*; hence the quartic surface corresponding to case (1) must be *syzygetic*. Moreover it will contain six nodes on a conic. If $D_2' \ldots D_6'$ are the intersections of the plane quintic and the line, then $D_1 \ldots D_6$ lie on a conic.

Two cases occur according as four or two of the associated nodes lie on this plane; in the first case since four of the associated nodes are coplanar, so also are the other four, and the equation of the surface is of the form

$$A^2 + \rho xyK = 0;$$

it is a case of XI_d. This surface is XII_d.

* A quartic surface through three generators of a quadric meets it also in a quintic; each generator of this set meets the quintic four times, hence (Salmon, p. 358) $H=3$ and therefore $h'=6$.

The surface has two tropes each containing six nodes: taking them as $D_1 \dots D_6$ and $D_2 D_3 D_7 \dots D_{10}$ it is clear that the tangent cones from the points $D_1 D_4 \dots D_{10}$ break up into a plane and a quintic cone; the tangent cones from D_2 and D_3 into two planes and a quartic cone; the tangent cone from D_{11} includes K, and therefore breaks up into a quadric cone and a quartic cone with three double edges.

When only two* of the points $D_1 \dots D_6$ are among the eight associated nodes, if e.g. they are D_1 and D_2, then the tangent cones from $D_3 \dots D_6$ break up into a quadric cone and a quartic cone, but this quartic cone must consist in part of the plane $D_1 \dots D_6$†, thus the tangent cone splits up into two cubic cones and we have a symmetroid with twelve nodes. This is XII_a.

The second case, a binodal quartic curve and two straight lines, leads in general to XII_d, i.e. $A^2 + \rho xyK = 0$, but if K breaks up into two planes we obtain the surface

$$A^2 + \rho xyzw = 0;$$

this is a twelve-nodal surface in which the tangent cone from each node breaks up into two planes and a quartic cone with two double edges. This surface is XII_c.

The cases (3) and (4) lead to the surface XII_a. Case (5) may lead to XII_d, but if in more than two cases the tangent cone from a node breaks up into a quadric cone and a trinodal quartic cone, we have a special case of XI_b. In this case *every* tangent cone must split up into such a quadric and quartic cone, otherwise we should obtain one of the preceding cases, which are excluded. The twelve nodes form three sets of eight associated points.

11. Quartic surfaces with thirteen nodes.

The plane sextics with twelve nodes divide themselves into the following classes:

(1) three conics,

(2) a nodal cubic, a conic and a straight line,

(3) a trinodal quartic and two straight lines,

(4) a cubic and three straight lines.

* The case in which three of the six points belong to the associated nodes cannot occur.

† The quadric cone cannot split up, as giving two tropes.

Three conics u, v, w with a common contact-conic form a degenerate sextic of the first kind arising as the projection of three conics u_1, v_1, w_1 upon the same quadric. The cone whose vertex is D_1 which stands on u meets this quadric in the pair of conics u_1, u_1'. Similarly we have the pair v_1, v_1' and w_1, w_1'; since u_1 and v_1 have two apparent points of intersection, the three conics u_1, v_1 and w_1 have six which lie on a quadric cone.

This applies also to the conics

$$u_1 v_1' w_1'; \quad u_1' v_1 w_1'; \quad u_1' v_1' w_1;$$

hence we have from the conics u, v, w four new conics upon which their twelve intersections lie by sixes. Hence there are *four* tropes*, and the surface is a case of XII_c, viz.

$$A^2 + \rho xyzw = 0.$$

The thirteenth node is one of the eight solutions of the equations

$$A_1 . x = A_2 . y = A_3 . z = A_4 . w.$$

The tangent cones at each of the first twelve nodes break up in each case into two planes and a quartic cone with three double edges. This surface is XIII_a.

The plane sextic (2) consisting of a line, a conic and a nodal cubic (all having a common contact-conic), is the projection of the complete intersection of a quadric and a cubic surface which have a line and a conic in common. Let D_2' be the node; $D_1' D_{11}' D_{12}'$ the intersections of the line and cubic; $D_3' D_4'$ the intersections of the line and conic; $D_5' \ldots D_{10}'$ those of the conic and cubic.

Considering the three loci on the quadric it is clear, since the generator meets the conic once and the cubic twice, while the conic and cubic meet three times, that there are five apparent intersections of these curves. Let their projections from D_{13} be $D_1' D_3' D_5' D_6' D_7'$; then these five points lie on a conic with D_2', hence $D_1 D_2 D_3 D_5 D_6 D_7$ lie on a conic.

The cone joining any point to the conic on the quadric meets the quadric in another conic; by associating this new conic with the generator and twisted cubic it is easily seen that the points

$$D_1 D_2 D_4 D_8 D_9 D_{10}$$

lie on a conic.

* See the first variety of surfaces with eleven nodes.

Hence we have that

$D_1D_2D_3D_5D_6D_7$ lie on a conic, let $x=0$ be its plane.

$D_1D_2D_4D_8D_9D_{10}$ .... $y=0$ ....

$D_1D_3D_4D_{11}D_{12}D_{13}$ .... $z=0$ ....

It follows that three tropes pass through D_1.

Since $D_1 \dots D_{10}$ lie on two conics intersecting in D_1 and D_2, a quadric S through $D_1 \dots D_{10}$ has an equation of the form

$$xu + yv - zw = 0;$$

hence since $x=0$, $y=0$ are tropes meeting the surface in two conics lying on S, the equation of the surface has the form

$$(xu + yv - zw)^2 + 4xyV = 0.$$

But since $z=0$ is also a trope it follows that $V \equiv zw' - uv$; hence the equation of the surface is

$$x^2u^2 + y^2v^2 + z^2w^2 - 2yzvw - 2zxwu - 2xyuv + 4xyzw' = 0.$$

This may be written in the form

$$\begin{vmatrix} 0 & z & y & u \\ z & 0 & x & v \\ y & x & 0 & w \\ u & v & w & w' \end{vmatrix} = 0.$$

This surface is XIII_b.

The tangent cone from D_1 consists of three planes and a cubic cone; the cones from $D_2D_3D_4$ of two planes and a quartic cone with three double edges; the cones from $D_5 \dots D_{13}$ of a plane, a quadric cone and a cubic cone with a double edge.

Hence if the plane sextic consists of three lines and a cubic we have XIII_b; if it consists of two lines and a trinodal quartic we have XIII_a or XIII_b.

12. Quartic surfaces with fourteen nodes.

The plane sextic with thirteen nodes is formed either by two lines and two conics or by three lines and a nodal cubic; it will be seen that either leads to the same fourteen-nodal quartic surface. For in XIII_a the tangent cone from one node splits up into three quadric cones; if there be another node one of these cones must consist of two planes $\alpha\beta$ which pass through the additional node D_{14}. Since $\alpha\beta$ is a tangent cone it will pass through eight associated points of the nodes $D_1 \dots D_{12}$; if the equation of the surface is $A^2 + \rho xyzw = 0$, let these eight points be taken as the intersections of

$$A = 0, \quad xy = 0, \quad zw = 0;$$

then
$$A \equiv \alpha\beta + pxy + qzw,$$

and the equation of the surface is

$$(\alpha\beta + pxy + qzw)^2 + \rho xyzw = 0.$$

But since $\alpha\beta$ is a pair of tropes it follows that $\rho = -4pq$, hence the surface is, with a slight change of notation,

$$x^2x'^2 + y^2y'^2 + z^2z'^2 - 2yzy'z' - 2zxz'x' - 2xyx'y' = 0,$$

or
$$\begin{vmatrix} 0 & z & y & x' \\ z & 0 & x & y' \\ y & x & 0 & z' \\ x' & y' & z' & 0 \end{vmatrix} = 0,$$

or again
$$\sqrt{xx'} + \sqrt{yy'} + \sqrt{zz'} = 0.$$

Also the pencil of surfaces included in XIII_b, viz.

$$x^2x'^2 + y^2y'^2 + z^2z'^2 - 2yzy'z' - 2zxz'x' - 2xyx'y' + \rho xyzw' = 0,$$

includes the preceding surface. Thus the addition of one node to XIII_a or to XIII_b leads to the same fourteen-nodal surface. It is to be observed that the surface has as tropes the planes

$$x = 0, \quad x' = 0, \quad y = 0, \quad y' = 0, \quad z = 0, \quad z' = 0,$$

and has as nodes the points

$$(xyz), \quad (xy'z), \quad (xyz'), \quad (x'yz), \quad (xy'z'), \quad (x'y'z), \quad (x'yz'), \quad (x'y'z') \qquad \dots\dots\dots(1),$$

together with the six points

$$x = x' = yy' - zz' = 0, \quad y = y' = zz' - xx' = 0, \quad z = z' = xx' - yy' = 0 \qquad \dots\dots\dots(2).$$

The tangent cone from any one of the first eight nodes consists of three planes and a cubic cone with a double edge; the tangent cone from either of the last six nodes consists of two planes and two quadric cones.

Surfaces with fifteen or with sixteen nodes.

Between the six planes $x \ldots z'$ there exists a singly infinite number of linear identities; if one of them is of the form

$$Ax + By + Cz + A'x' + B'y' + C'z' \equiv 0 \quad \ldots\ldots\ldots(3),$$

with the condition

$$AA' = BB' = CC',$$

then one condition is imposed upon the system of six planes; and if this condition is satisfied the point given by the equations

$$Ax = A'x', \quad By = B'y', \quad Cz = C'z'$$

is a node of the surface. For this point is seen to lie on the surface, and at this point the differential equation

$$\frac{x\,dx' + x'dx}{\sqrt{xx'}} + \frac{y\,dy' + y'dy}{\sqrt{yy'}} + \frac{z\,dz' + z'dz}{\sqrt{zz'}} = 0,$$

giving consecutive points on the tangent plane thereat, becomes

$$\frac{dx'}{A} + \frac{dx}{A'} + \frac{dy'}{B} + \frac{dy}{B'} + \frac{dz'}{C} + \frac{dz}{C'} = 0,$$

which vanishes identically, as is seen by differentiating the equation (3) and using the condition $AA' = BB' = CC'$. Hence this point is a node of the surface. The surface is therefore fifteen-nodal if this condition is satisfied.

The sextic cone from any node now splits up into four planes and a quadric cone. There are ten tropes, viz. the planes $x \ldots z'$ and the four planes

$$Cz + Ax + B'y' = 0, \qquad Cz + A'x' + By = 0,$$
$$C'z' + Ax + By = 0, \qquad C'z' + A'x' + B'y' = 0.$$

These planes are seen to be tropes since, for instance, the plane

$$A'x' + B'y' + C'z' = 0$$

passes through the fifteenth node, the points (xyz), $(x'y'z')$, and through one of each of the three pairs of nodes (2); it thus contains six nodes and is therefore a trope.

If a second linear identity between the planes $x \ldots z'$ exists, the constants of which are connected by a similar equation, there will be a sixteenth node*, and we have the sixteen-nodal surface of Kummer.

* See also a paper by the author, *Quarterly Journal of Mathematics*, 1900.

The surfaces which have been discussed in this chapter are the following:

Surface	Number of nodes
$(a \between A, B, C, D, E, F)^2$	4
$(a \between A, B, C, D, E)^2$	5
$(a \between A, B, C, D)^2 + \rho J$	6
$(a \between A, B, C)^2 + \rho \Sigma$	7
$(a \between A, B, C)^2$	8
$(a \between A, B)^2 + \rho T$	8
$A^2 + \rho KB$	9
K being a cone whose vertex is on A	
$A^2 + \rho P$	9
Thirteen of the pencil of surfaces $A^2 + \rho P$ where P is any quartic surface with nine nodes and A passes through them; *one* of these surfaces is a symmetroid	10
$A^2 + \rho xyB$	10
$A^2 + \rho K_1 K_2$	10
where K_1 and K_2 are cones whose vertices lie on A	
Symmetroid with eleven nodes, XI_a	11
$Kw^2 + 2wu_3 + u_2^2$, XI_c	11
$A^2 + \rho K_1 K_2$, XI_b,	11
a case of preceding,	
$A^2 + \rho Kxy$, XI_d	11
$A^2 + \rho xyzw$, XII_c	12
$A^2 + \rho K_1 K_2$, XII_b,	12
a case of preceding,	
Symmetroid with twelve nodes, XII_a	12
$A^2 + \rho Kxy$, XII_d	12
$A^2 + \rho xyzw$	13
$x^2u^2 + y^2v^2 + z^2w^2 - 2yzvw - 2zxwu - 2xyuv + 4xyzw'$	13
$\sqrt{xx'} + \sqrt{yy'} + \sqrt{zz'}$	14
The same, where $Ax + By + Cz + A'x' + B'y' + C'z' \equiv 0$, with the condition $AA' = BB' = CC'$	15
The same, where an additional condition of this form exists.	16

CHAPTER II

DESMIC SURFACES

13. An interesting type of quartic surface which possesses nodes but not singular curves is afforded by desmic surfaces. Desmic* surfaces are such that a pencil of such surfaces contains the special quartics formed by three tetrahedra. The equation of a desmic surface is

$$\lambda\Delta_1 + \mu\Delta_2 + \nu\Delta_3 = 0, \text{ where } \Delta_1 = \prod_1^4 (a_i x_1 + a_i' x_2 + a_i'' x_3 + a_i''' x_4), \text{ etc.,}$$

and where an identity exists of the form

$$\alpha\Delta_1 + \beta\Delta_2 + \gamma\Delta_3 \equiv 0.$$

Such tetrahedra are called desmic. They are shown to exist by consideration of such an identity as

$$(x^2 - y^2)(z^2 - t^2) + (x^2 - t^2)(y^2 - z^2) + (x^2 - z^2)(t^2 - y^2) = 0 \quad \ldots(1).$$

Writing the preceding identity in the form

$$\prod_1^4 A_i + \prod_1^4 B_i + \prod_1^4 C_i \equiv 0,$$

it is clear that any face of Δ_1 and any face of Δ_2 are coaxal with some face of Δ_3. Hence Δ_2 may be written in any one of the forms

$$\Pi(C_1 + \kappa_i A_i), \quad \Pi(C_2 + \kappa_i' A_i), \quad \Pi(C_3 + \kappa_i'' A_i), \quad \Pi(C_4 + \kappa_i''' A_i).$$

It follows that the edge (A_1, A_2) of Δ_1 meets Δ_2 in *edges* of the latter, viz. at the points $A_1 = A_2 = C_i = 0$, $(i = 1, 2, 3, 4)$, and *two* of these points are necessarily distinct since the faces of Δ_3 are not concurrent. These two edges of Δ_2 do not intersect, for otherwise (A_1, A_2) would lie in a face of Δ_2. Also since $\Delta_2 \equiv \Pi(C_1 + \kappa_i A_i)$ it is clear that (A_1, A_2) and (A_3, A_4) meet opposite edges of Δ_2, i.e. (A_1, A_2) and (A_3, A_4) meet the *same pair* of opposite edges of Δ_2. Hence *any pair of non-intersecting edges of one tetrahedron meet a pair of non-intersecting edges of either of the other two*

* δεσμός = pencil. See Humbert, *Sur les surfaces desmiques*, Liouville (1891).

tetrahedra; also we obtain sixteen lines through each of which a face of each tetrahedron passes.

Taking Δ_1 as tetrahedron of reference and one face of Δ_2 as $x+y+z+t=0$, the identity becomes

$$xyzt + (x+y+z+t)\,B_2B_3B_4 + C_1C_2C_3C_4 = 0;$$

and the fact that any two opposite edges of Δ_1 meet two opposite edges of Δ_2 leads at once to the form of B_2, B_3 and B_4. Finally the identity

$$16xyzt - (x+y+z+t)(x+y-z-t)(x-y+z-t)(x-y-z+t)$$
$$-(x+y+z-t)(x+y-z+t)(x-y+z+t)(-x+y+z+t) = 0 \quad \text{......(2)}$$

shows the form of $C_1 \ldots C_4$.

The form obtained for Δ_2 shows that any edge, e.g. (xy), of Δ_1 meets opposite edges of Δ_2 in two points harmonic with the points (xyz), (xyt), hence *any two vertices of Δ_1 are harmonic with the points in which their join meets opposite edges of Δ_2.* Hence if Δ_1 is given, a tetrahedron Δ_2 desmic with it is obtained as follows: *if P be any point, draw through P a line to intersect a pair of opposite edges of Δ_1 and let P' be the fourth harmonic to P and the points of intersection, also let P'', P''' be the two other points similarly determined, then the tetrahedron $PP'P''P'''$ is desmic to Δ_1.*

The identity

$$4(x^2+y^2+z^2+t^2)$$
$$=(x+y+z+t)^2+(x+y-z-t)^2+(x-y+z-t)^2+(x-y-z+t)^2$$
$$=(x+y+z-t)^2+(x+y-z+t)^2+(x-y+z+t)^2+(-x+y+z+t)^2$$

shows that Δ_1, Δ_2, Δ_3 are self-polar for the quadric

$$x^2+y^2+z^2+t^2=0.$$

Hence since the intersection of any two faces of Δ_2 and Δ_3 lies in a face of Δ_1, it follows by reciprocation that the join of any two vertices of Δ_2 and Δ_3 passes through a vertex of Δ_1; we thus obtain sixteen lines each of which contains a vertex of each tetrahedron. Therefore three desmic tetrahedra are such that any pair of them have four centres of perspective, viz. the vertices of the third tetrahedron. Conversely if two tetrahedra have four centres of perspective they are in desmic position. For it is easy to see that two such tetrahedra have the property that each pair of opposite edges of one tetrahedron meets a pair of opposite edges of the other tetrahedron, and this necessarily involves that the tetrahedra are in desmic position, as may be seen by expressing the latter conditions.

The identity (1) affords another system of desmic tetrahedra D_1, D_2, D_3 closely related to that given by (2): the faces of D_1, D_2, D_3 are respectively

$$x-y=0,\quad x+y=0,\quad z-t=0,\quad z+t=0;$$
$$x-z=0,\quad x+z=0,\quad y-t=0,\quad y+t=0;$$
$$x-t=0,\quad x+t=0,\quad y-z=0,\quad y+z=0.$$

The vertices of the three tetrahedra Δ_i which arise from (2) being respectively

I	(0001)	(0010)	(0100)	(1000)
II	(1111)	$(11\bar{1}\bar{1})$	$(1\bar{1}1\bar{1})$	$(1\bar{1}\bar{1}1)$
III	$(111\bar{1})$	$(11\bar{1}1)$	$(1\bar{1}11)$	$(\bar{1}111)$

it may be observed that the preceding sixteen lines joining the vertices of Δ_i are the intersections of the faces of two tetrahedra D_i (e.g. the planes $x-y=0$, $x-z=0$ contain the three points in the first column); and that the join of two vertices of a Δ meets two opposite edges of another Δ in two vertices of a D.

14. Desmic surfaces.

We may therefore take as the equation of the general desmic surface the equation

$$aD_1+bD_2+cD_3=0,$$

where
$$D_1+D_2+D_3\equiv 0.$$

This may be written in the form

$$(x^2-y^2)(z^2-t^2)+k(x^2-z^2)(y^2-t^2)=0;$$

it has the twelve points I, II, III, the vertices of the Δ_i, as nodes; and contains each of the sixteen lines joining the vertices of Δ_i by threes.

The equation of the surface may be written in the form

$$\frac{\lambda}{\alpha}+\frac{\mu}{\beta}+\frac{\nu}{\gamma}=0,\quad \lambda+\mu+\nu=0,$$

where
$$\alpha=x^2-y^2,\quad \beta=x^2-z^2,\quad \gamma=x^2-t^2.$$

Now any quadric through the eight points II, III is clearly

$$A\alpha+B\beta+C\gamma=0,$$

whence it follows that this quadric meets the desmic surface in two quadri-quartics; and that these curves form a simply infinite system.

Again the desmic surface may be written

$$\lambda(x^2y^2+z^2t^2)+\mu(x^2z^2+y^2t^2)+\nu(x^2t^2+y^2z^2)=0, \quad \lambda+\mu+\nu=0;$$

or
$$\lambda(xy-zt)^2+\mu(xz-yt)^2+\nu(xt-yz)^2=0.$$

This surface is intersected by the quadrics

$$A(xy-zt)+B(xz-yt)+C(xt-yz)=0$$

in pairs of quadri-quartics. These quadrics pass through the points I, II. Similar considerations apply to the quadrics

$$A(xy+zt)+B(xz+yt)+C(xt+yz)=0,$$

which are those passing through the points I, III. Hence there exist three systems of quadri-quartics on the surface. Through each point of the surface there passes one curve of each system.

It is known that the generators of the system

$$A\alpha+B\beta+C\gamma=0,$$

as belonging to the quadrics through eight associated points, form a cubic complex. The quadrics contain four systems of cones having their vertices at the points I, and any line through any one of these four points belongs to one cone of its system. Hence every line through the points I belongs to the cubic complex, and it is clear that every line through the points II and III belongs to this complex, which is thus determined by the twelve points I, II, III (for the join of any point P to these points gives twelve lines of the complex through P). The complex is therefore the same whichever system of quadrics be used.

Considering any line p of this complex, p is thus a generator of a quadric of each of the three systems, hence it is a chord of each of three pairs of quadri-quartics: thus if p meets the desmic surface in the points a_1, a_2, a_3, a_4, then

a_1a_2 and a_3a_4 belong to a quadri-quartic of the system I, II,
a_1a_3 and a_2a_4 .... I, III,
a_1a_4 and a_2a_3 .... II, III.

15. Expression of the surface in terms of σ functions.

Consider the surface defined by the equations

$$\rho.x=\frac{\sigma_1(u)}{\sigma_1(v)}, \quad \rho.y=\frac{\sigma_2(u)}{\sigma_2(v)}, \quad \rho.z=\frac{\sigma_3(u)}{\sigma_3(v)}, \quad \rho.t=\frac{\sigma(u)}{\sigma(v)};$$

so that $\frac{x}{t} = \frac{\sigma_1(u)}{\sigma(u)} \div \frac{\sigma_1(v)}{\sigma(v)}$, with similar expressions for $\frac{y}{t}, \frac{z}{t}$, where the functions σ are defined as follows:

$$\left(\frac{\sigma_1(u)}{\sigma(u)}\right)^2 = \wp(u) + e_1, \text{ etc.}$$

It follows that

$$\frac{x^2}{x^2 - t^2} = \frac{\wp(u) + e_1}{\wp(u) - \wp(v)}, \quad \frac{y^2}{y^2 - t^2} = \frac{\wp(u) + e_2}{\wp(u) - \wp(v)},$$

$$\frac{z^2}{z^2 - t^2} = \frac{\wp(u) + e_3}{\wp(u) - \wp(v)},$$

whence eliminating $\wp(u)$, $\wp(v)$ we obtain

$$\begin{vmatrix} 1 & x^2 & e_1(x^2 - t^2) \\ 1 & y^2 & e_2(y^2 - t^2) \\ 1 & z^2 & e_3(z^2 - t^2) \end{vmatrix} = 0.$$

This gives on expansion

$$(e_1 - e_2)(x^2y^2 + z^2t^2) + (e_3 - e_1)(x^2z^2 + y^2t^2) + (e_2 - e_3)(x^2t^2 + y^2z^2) = 0;$$

which is the form of the equation of the desmic surface previously obtained.

Nodes, lines and quadri-quartics of the surface.

If $2\omega_1$, $2\omega_2$ are the periods of $\wp(\theta)$ and if $\omega_1 + \omega_2 + \omega_3 = 0$, then since

$$\frac{\sigma_1(\theta + 2\omega_1)}{\sigma(\theta + 2\omega_1)} = \frac{\sigma_1(\theta)}{\sigma(\theta)}, \quad \frac{\sigma_1(\theta + 2\omega_2)}{\sigma(\theta + 2\omega_2)} = -\frac{\sigma_1(\theta)}{\sigma(\theta)}, \text{ etc.},$$

it follows by considering the ratios $\frac{x}{t}, \frac{y}{t}, \frac{z}{t}$, that to any point of the surface there corresponds an infinite number of arguments of the form

$$\epsilon u + 2k\omega_1 + 2k'\omega_2 + 4h\omega_1 + 4h'\omega_2,$$

$$\epsilon v + 2k\omega_1 + 2k'\omega_2 + 4h_1\omega_1 + 4h_1'\omega_2; \quad \epsilon = \pm 1.$$

We obtain the nodes I when v has the values $0, \omega_1, \omega_2, \omega_3$,

.... II ...... $u - v$.............. $0, 2\omega_1, 2\omega_2, 2\omega_3$,

.... III...... $u + v$.............. $0, 2\omega_1, 2\omega_2, 2\omega_3$.

The sixteen lines of the surface correspond to the equations

$u = 0, \quad v = 2k\omega_1 + 2k'\omega_2;$ $\qquad u = \omega_1, \quad v = \omega_1 + 2k\omega_1 + 2k'\omega_2;$

$u = \omega_2, \quad v = \omega_2 + 2k\omega_1 + 2k'\omega_2;$ $\qquad u = \omega_3, \quad v = \omega_3 + 2k\omega_1 + 2k'\omega_2;$

where k, k' are zero or unity.

The well-known relations*

$$\sigma(u+v)\,\sigma(u-v)=\sigma^2(u)\,\sigma_\lambda{}^2(v)-\sigma^2(v)\,\sigma_\lambda{}^2(u),$$

$$\sigma_\lambda(u-v)\,\sigma(u+v)=\sigma_\mu(u)\,\sigma_\nu(u)\,\sigma_\lambda(v)\,\sigma(v)+\sigma_\mu(v)\,\sigma_\nu(v)\,\sigma_\lambda(u)\sigma(u),$$

$$\sigma_\lambda(u+v)\,\sigma(u-v)=\sigma_\lambda(u)\,\sigma(u)\,\sigma_\mu(v)\,\sigma_\nu(v)-\sigma_\mu(u)\,\sigma_\nu(u)\,\sigma_\lambda(v)\sigma(v),$$

lead to the following identities when the values of x, y, z, t are inserted, viz.

$$A(x^2-t^2)+B(y^2-t^2)+C(z^2-t^2)$$
$$\equiv-\frac{\sigma(u+v)\,\sigma(u-v)}{\rho^2\sigma^2(v)}\left\{\frac{A}{\sigma_1{}^2(v)}+\frac{B}{\sigma_2{}^2(v)}+\frac{C}{\sigma_3{}^2(v)}\right\},$$

$$A(xy+zt)+B(xz+yt)+C(yz+xt)$$
$$=\frac{\sigma(u+v)\,\{A\sigma_3(u-v)+B\sigma_2(u-v)+C\sigma_1(u-v)\}}{\rho^2\,\sigma(v)\,\sigma_1(v)\,\sigma_2(v)\,\sigma_3(v)},$$

$$A(xy-zt)+B(xz-yt)+C(yz-xt)$$
$$=-\frac{\sigma(u-v)\,\{A\sigma_3(u+v)+B\sigma_2(u+v)+C\sigma_1(u+v)\}}{\rho^2\sigma(v)\,\sigma_1(v)\,\sigma_2(v)\,\sigma_3(v)}.$$

Hence it follows that for the quadri-quartics II, III

$$v=\text{constant};$$

for the quadri-quartics I, III

$$u-v=\text{constant};$$

and for the quadri-quartics I, II

$$u+v=\text{constant}.$$

It is to be observed that the curve $v=\alpha$ is identical with the curve

$$v=\pm\alpha+2hw_1+2h'w_2;$$

and that $u-v=\alpha$ is identical with

$$u-v=\pm\alpha+4hw_1+4h'w_2.$$

16. Intersection of a line of the cubic complex with the surface.

Any line p of the preceding cubic complex is a chord of three pairs of quadri-quartics: if $(u_1v_1)\ldots(u_4v_4)$ are the arguments of its four points of intersection with the surface, let a pair of curves of the system II, III be $v=\alpha$, $v=\beta$, then we may take

$$v_1=v_4=\alpha,\qquad v_2=v_3=\beta;$$

* See Harkness and Morley, *Theory of Functions*, p. 315.

the u_i are then connected by the equations

$$u_1 + \alpha = \epsilon_1 (u_2 + \beta), \quad u_4 + \alpha = \epsilon_2 (u_3 + \beta),$$
$$u_1 - \alpha = \epsilon_3 (u_3 - \beta), \quad u_4 - \alpha = \epsilon_4 (u_2 - \beta),$$

where $\epsilon_i = \pm 1$.

Since p is any generator of the quadric containing the curves $v = \alpha$, $v = \beta$, the u_i each involve one indeterminate, hence on subtracting the third equation from the first and the fourth from the second and identifying the results we obtain

$$\epsilon_1 = \epsilon_2, \quad \epsilon_3 = \epsilon_4, \quad \epsilon_1 = -\epsilon_4;$$

hence taking $\epsilon_1 = 1$* and writing $u_1 = \beta + \mu$, the arguments of the points of intersection are given as

$$u_1 = \beta + \mu, \quad u_4 = \beta - \mu, \quad u_2 = \alpha + \mu, \quad u_3 = \alpha - \mu;$$
$$v_1 = \alpha, \quad v_4 = \alpha, \quad v_2 = \beta, \quad v_3 = \beta.$$

17. Bitangents of the surface.

The tangents to the quadri-quartics of the three systems which pass through the point (u, v) are bitangents of the surface; for

$$(u_1, v_1) \equiv (u_4, v_4) \text{ if } \mu = 0 \quad \text{and then } (u_2, v_2) \equiv (u_3, v_3);$$
$$(u_1, v_1) \equiv (u_2, v_2) \text{ if } \alpha = \beta \quad \ldots\ldots \quad (u_3, v_3) \equiv (u_4, v_4);$$
$$(u_1, v_1) \equiv (u_3, v_3) \text{ if } \alpha = -\beta \quad \ldots\ldots \quad (u_2, v_2) \equiv (u_4, v_4).$$

It also follows that the three bitangents of the surface determined by the point (u, v) touch it at the points (v, u), $(2v - u, v)$, $(-2v - u, v)$. These three points are *collinear*, since the join of the points $(2v - u, v)$, $(-2v - u, v)$ by the preceding Article meets the surface in the points (v, u), $(-3v, u)$.

If p touches curves of the system II, III at P and Q so that P is the point (α, β) and Q the point (β, α), then, as Q moves to a consecutive position on the curve $v = \alpha$, P takes a consecutive position on the curve $u = \alpha$; thus the tangent plane to the surface at P' passes through PQ, so that the tangents at P to the curves $u = \alpha$, $v = \beta$ are *conjugate*, and the curves $u = \text{constant}$, $v = \text{constant}$ form a conjugate network on the surface.

Similarly it is seen that the system conjugate to $u + v = \text{const.}$ is $3v - u = \text{const.}$, and that the system conjugate to $u - v = \text{const.}$ is $3v + u = \text{const.}$

We can now determine the relation connecting any pair of conjugate tangents at any given point of the surface; for if du, dv;

* Taking $\epsilon_1 = -1$ gives the same form of result.

du_1, dv_1 correspond to this pair of conjugate tangents we have an involutive equation of the form

$$du\,du_1 + p\,(du\,dv_1 + du_1 dv) + q\,dv\,dv_1 = 0,$$

where p and q are functions of u and v.

Expressing that this equation is satisfied by $du = dv_1 = 0$ and by $du - dv = 3dv_1 + du_1 = 0$, we obtain that

$$p = 0, \quad q = 3;$$

and the equation assumes the form

$$du\,du_1 + 3dv\,dv_1 = 0.$$

The asymptotic lines correspond to the assumption $du = du_1$, $dv = dv_1$ and their differential equation is therefore

$$du^2 + 3dv^2 = 0,$$

whose integrated form is

$$u + \sqrt{3}iv = \text{constant}, \quad u - \sqrt{3}iv = \text{constant}.$$

18. Plane sections of the surface.

The plane

$$(e_2 - e_3)\,\sigma_1(\alpha)\,\sigma_1(\beta)\,\sigma_1(\gamma)\,\sigma_1(\delta)\,x + \ldots + \ldots$$
$$- (e_1 - e_2)(e_2 - e_3)(e_3 - e_1)\,\sigma(\alpha)\,\sigma(\beta)\,\sigma(\gamma)\,\sigma(\delta)\,t = 0$$

passes through the sixteen points whose arguments are

$$\begin{array}{llll} \beta+\gamma+\delta,\ \alpha; & \beta-\gamma-\delta,\ \alpha; & \alpha+\gamma+\delta,\ \beta; & \alpha-\gamma-\delta,\ \beta; \\ \gamma-\delta-\beta,\ \alpha; & \delta-\gamma-\beta,\ \alpha; & \delta-\alpha-\gamma,\ \beta; & \gamma-\delta-\alpha,\ \beta; \\ \alpha+\beta+\delta,\ \gamma; & \delta-\alpha-\beta,\ \gamma; & \alpha+\beta+\gamma,\ \delta; & \gamma-\alpha-\beta,\ \delta; \\ \alpha-\beta-\delta,\ \gamma; & \beta-\delta-\alpha,\ \gamma; & \beta-\gamma-\alpha,\ \delta; & \alpha-\beta-\gamma,\ \delta. \end{array}$$

For two forms of the "equation of three terms" of the σ-functions are*

$$(e_\mu - e_\nu)\,\sigma_\lambda(a)\,\sigma_\lambda(b)\,\sigma_\lambda(c)\,\sigma_\lambda(d) + (e_\nu - e_\lambda)\,\sigma_\mu(a')\,\sigma_\mu(b')\,\sigma_\mu(c')\,\sigma_\mu(d')$$
$$+ (e_\lambda - e_\mu)\,\sigma_\nu(a'')\,\sigma_\nu(b'')\,\sigma_\nu(c'')\,\sigma_\nu(d'') = 0;$$

$$\sigma_\lambda(a)\,\sigma_\lambda(b)\,\sigma_\lambda(c)\,\sigma_\lambda(d) - \sigma_\lambda(a')\,\sigma_\lambda(b')\,\sigma_\lambda(c')\,\sigma_\lambda(d')$$
$$+ (e_\lambda - e_\mu)(e_\lambda - e_\nu)\,\sigma(a'')\,\sigma(b'')\,\sigma(c'')\,\sigma(d'') = 0;$$

where

$$\begin{aligned} 2a' &= a+b+c+d, & 2a'' &= a+b+c-d, \\ 2b' &= a+b-c-d, & 2b'' &= a+b-c+d, \\ 2c' &= a-b+c-d, & 2c'' &= a-b+c+d, \\ 2d' &= -a+b+c-d, & 2d'' &= a-b-c-d. \end{aligned}$$

* See Harkness and Morley, *Theory of Functions*, p. 313.

In the first formula taking

$$\begin{array}{llll} a = 0, & b = \alpha+\beta, & c = \alpha+\gamma, & d = \beta+\gamma, \\ a' = \alpha+\beta+\gamma, & b' = -\gamma, & c' = -\beta, & d' = \alpha, \\ a'' = \alpha, & b'' = \beta, & c'' = \gamma, & d'' = -(\alpha+\beta+\gamma), \end{array}$$

we obtain

$$\begin{aligned}(e_\mu - e_\nu)\,\sigma_\lambda(\alpha+\beta)\,\sigma_\lambda(\alpha+\gamma)\,\sigma_\lambda(\beta+\gamma) & \\ + (e_\nu - e_\lambda)\,\sigma_\mu(\alpha+\beta+\gamma)\,\sigma_\mu(\alpha)\,\sigma_\mu(\beta)\,\sigma_\mu(\gamma) & \\ + (e_\lambda - e_\mu)\,\sigma_\nu(\alpha+\beta+\gamma)\,\sigma_\nu(\alpha)\,\sigma_\nu(\beta)\,\sigma_\nu(\gamma) &= 0.\end{aligned}$$

In the second formula let

$$\begin{array}{llll} a = \alpha+\beta+\gamma, & b = -\alpha, & c = -\beta, & d = -\gamma, \\ a' = 0, & b' = \beta+\gamma, & c' = \gamma+\alpha, & d' = -(\alpha+\beta), \\ a'' = \gamma, & b'' = \beta, & c'' = \alpha, & d'' = \alpha+\beta+\gamma, \end{array}$$

we then obtain

$$\begin{aligned}\sigma_\lambda(\alpha+\beta+\gamma)\,\sigma_\lambda(\alpha)\,\sigma_\lambda(\beta)\,\sigma_\lambda(\gamma) - \sigma_\lambda(\beta+\gamma)\,\sigma_\lambda(\gamma+\alpha)\,\sigma_\lambda(\alpha+\beta) & \\ + (e_\lambda - e_\mu)(e_\lambda - e_\nu)\,\sigma(\alpha)\,\sigma(\beta)\,\sigma(\gamma)\,\sigma(\alpha+\beta+\gamma) &= 0.\end{aligned}$$

On substitution from the previous result it follows that

$$\begin{aligned}(e_\mu - e_\nu)\,\sigma_\lambda(\alpha+\beta+\gamma)\,\sigma_\lambda(\alpha)\,\sigma_\lambda(\beta)\,\sigma_\lambda(\gamma) & \\ + (e_\nu - e_\lambda)\,\sigma_\mu(\alpha+\beta+\gamma)\,\sigma_\mu(\alpha)\,\sigma_\mu(\beta)\,\sigma_\mu(\gamma) & \\ + (e_\lambda - e_\mu)\,\sigma_\nu(\alpha+\beta+\gamma)\,\sigma_\nu(\alpha)\,\sigma_\nu(\beta)\,\sigma_\nu(\gamma) & \\ - (e_\lambda - e_\mu)(e_\mu - e_\nu)(e_\nu - e_\lambda)\,\sigma(\alpha+\beta+\gamma)\,\sigma(\alpha)\,\sigma(\beta)\,\sigma(\gamma) &= 0.\end{aligned}$$

This shows that the preceding plane passes through the point

$$x = \frac{\sigma_1(\alpha+\beta+\gamma)}{\sigma_1(\delta)}, \quad y = \frac{\sigma_2(\alpha+\beta+\gamma)}{\sigma_2(\delta)}, \quad z = \frac{\sigma_3(\alpha+\beta+\gamma)}{\sigma_3(\delta)},$$

$$t = \frac{\sigma(\alpha+\beta+\gamma)}{\sigma(\delta)}.$$

Since the function $\sigma_i(u)$ is an even function, it follows that the plane passes through the four points in the above table for which $v = \delta$; similarly it must pass through the other twelve points.

The fact that these sixteen points are coplanar may also be seen as follows*: denote the points by the notation a_{ik}, where the first suffix relates to the *row* and the second to the *column*: on comparing with the arguments of the four points on a line of the cubic complex, it follows that the four points a_{i1}, a_{i2}, a_{i3}, a_{i4}

* See Humbert, *loc. cit.*

are on such a line and also the points $a_{1i}, a_{2i}, a_{3i}, a_{4i}$; and four lines are given by the following groups of four points, viz.

$$a_{14},\ a_{23},\ a_{32},\ a_{41},$$
$$a_{13},\ a_{24},\ a_{31},\ a_{42},$$
$$a_{12},\ a_{21},\ a_{34},\ a_{43},$$
$$a_{11},\ a_{22},\ a_{33},\ a_{44}.$$

Also the four points $a_{11}, a_{12}, a_{21}, a_{22}$ are coplanar, since they have the same argument for v and the sum of the arguments for u is zero*.

It follows that the sixteen points are coplanar. They lie upon three sets of four lines of the cubic complex. Varying α we obtain an infinite number of such sets of sixteen points on any given plane.

The three systems of four lines touch a curve of the third class, to each of its tangents there corresponds an elliptic argument of periods Ω, Ω' say. Let the three sets of four lines have arguments a_i, b_i and c_i respectively, then expressing that through each point a_{ik} there pass three lines, one of each set, we have

$$a_1 + b_1 + c_4 \equiv 0,$$
$$a_1 + b_2 + c_3 \equiv 0,$$
$$a_1 + b_3 + c_2 \equiv 0,$$
$$a_1 + b_4 + c_1 \equiv 0,$$

* Consider the determinant

$$\begin{vmatrix} \frac{\sigma_1(u_1)}{\sigma(u_1)}, & \frac{\sigma_2(u_1)}{\sigma(u_1)}, & \frac{\sigma_3(u_1)}{\sigma(u_1)}, & 1 \\ \frac{\sigma_1(u_2)}{\sigma(u_2)}, & \cdots & \cdots & 1 \\ \cdots & \cdots & \cdots & \cdots \\ \frac{\sigma_1(u_4)}{\sigma(u_4)}, & \cdots & \cdots & 1 \end{vmatrix},$$

regarded as function of u_1 it is doubly periodic with periods $4\omega_1$, $4\omega_2$, and has four poles in a parallelogram of periods which are congruent to

$$0,\ 2\omega_1,\ 2\omega_2,\ 2\omega_1 + 2\omega_2;$$

hence it has four zeros congruent to

$$u_2,\ u_3,\ u_4,\ -(u_2+u_3+u_4).$$

Hence, expressed in terms of σ-functions, the determinant has a factor $\sigma(u_1+u_2+u_3+u_4)$. Therefore, if $\sum_1^4 u_i = 0$, four points for which the v is the same are coplanar.

$$a_2+b_1+c_3\equiv 0,\quad a_3+b_1+c_2\equiv 0,\quad a_4+b_1+c_1\equiv 0,$$
$$a_2+b_2+c_4\equiv 0,\quad a_3+b_2+c_1\equiv 0,\quad a_4+b_2+c_2\equiv 0,$$
$$a_2+b_3+c_1\equiv 0,\quad a_3+b_3+c_4\equiv 0,\quad a_4+b_3+c_3\equiv 0,$$
$$a_2+b_4+c_2\equiv 0,\quad a_3+b_4+c_3\equiv 0,\quad a_4+b_4+c_4\equiv 0,$$

whence we deduce that

$$2a_1\equiv 2a_2\equiv 2a_3\equiv 2a_4,\quad a_1+a_4\equiv a_2+a_3,$$
$$2b_1\equiv 2b_2\equiv 2b_3\equiv 2b_4,\quad b_1+b_4\equiv b_2+b_3,$$
$$2c_1\equiv 2c_2\equiv 2c_3\equiv 2c_4,\quad c_1+c_4\equiv c_2+c_3.$$

The solution of these equations is seen to be

$$a_1=-a,\quad a_2=-a+\frac{\Omega}{2},\quad a_3=-a+\frac{\Omega'}{2},\quad a_4=-a+\frac{\Omega}{2}+\frac{\Omega'}{2};$$

$$b_1=-b,\quad b_2=-b+\frac{\Omega}{2},\quad b_3=-b+\frac{\Omega'}{2},\quad b_4=-b+\frac{\Omega}{2}+\frac{\Omega'}{2};$$

$$c_1=-c+\frac{\Omega}{2}+\frac{\Omega'}{2},\quad c_2=-c+\frac{\Omega'}{2},\quad c_3=-c+\frac{\Omega}{2},\quad c_4=-c;$$

with the condition $a+b+c\equiv 0$.

Now the arguments of the lines a_i are seen to be those of the four tangents at the points in which the tangent of argument $2a$ meets the curve*; similarly for the lines b_i, c_i, hence we have the result that if C is the curve of the third class which is touched by the twelve lines a_i, b_i, c_i, *the three sets of four points of contact with C of the lines a_i, b_i and c_i lie on three tangents to C which are concurrent.*

Hence we derive a desmic configuration as follows. From any point P of the plane draw three tangents to a given curve C of the third class; each tangent meets C in four points in addition to its points of contact; the tangents at these points give rise to a desmic configuration of sixteen points Q; conversely if C and one of the points Q are given, the point P is uniquely determined.

If P describes a straight line the points Q describe a curve K of order μ; and since two different points P cannot give rise to the same point Q, two curves K can only have in common the sixteen points Q arising from the point of intersection of the two lines which give rise to these curves; hence $\mu^2=16$, i.e. K is of the fourth order.

* Clebsch, *Vorlesungen über Geometrie*, p. 607.

Conversely, every quartic curve through the sixteen points of a configuration can be generated in this manner. For if P be the point of the plane which corresponds to the given sixteen points, then one line through P can be chosen such that the quartic curve deduced from the line by this method meets the given quartic curve in any assigned point of the latter; the two quartics hence intersect in seventeen points and are therefore identical.

19. Sections by tangent planes.

We now consider the form of the section of the surface by a plane which touches the surface at any point P.

From P six tangents can be drawn to touch the curve of section; it has been seen that the points of contact of three of these tangents are collinear, viz. the tangents to the three quadri-quartics through P.

Hence the curve must have an equation of the form

$$z^2xy + \alpha\beta\gamma\,(ax + by + cz) = 0,$$

where the inflexional tangents at P and the line joining the points of contact of the above three tangents form the triangle of reference. If (xyz) is a point near P, we may (Art. 17) take

$$x = \delta u - i\sqrt{3}\,\delta v, \quad y = \delta u + i\sqrt{3}\,\delta v,$$

and, since the directions of α, β, γ are respectively given by

$$\delta v = 0, \quad \delta u - \delta v = 0, \quad \delta u + \delta v = 0,$$

it follows that

$$\alpha\beta\gamma \equiv (x - y)(x - \omega y)(x - \omega^2 y), \quad \omega^3 = 1.$$

Hence the equation of the curve of section by a tangent plane is

$$z^2xy + (x^3 - y^3)(ax + by + cz) = 0^*.$$

20. If p, q, r are three lines of a cubic surface, forming a triangle, any three planes through p, q, r respectively meet the surface also in conics which lie on a quadric. Cremona has shown that the locus of the vertices of such of these quadrics as degenerate into cones is a desmic surface. This will now be proved.

* The points of contact of the other three tangents from P to the curve are seen to lie on the line $ax + by + \frac{c}{2}z = 0$.

For let the cubic surface be

$$ft + xyz = 0,$$

then, if $x' = x - \alpha t,\quad y' = y - \beta t,\quad z' = z - \gamma t,$

the equation of the surface may be written

$$t\{f + \alpha yz + \beta zx + \gamma xy - \alpha\beta zt - \alpha\gamma yt - \beta\gamma xt + \alpha\beta\gamma t^2\} + x'y'z' = 0.$$

Denoting by F the coefficient of t, $F=0$ is the quadric which contains the three conics; if F is a cone the coordinates of its vertex are given by the equations

$$f_x + \beta z + \gamma y - \beta\gamma t = 0$$
$$f_y + \alpha z + \gamma x - \alpha\gamma t = 0,$$
$$f_z + \beta x + \alpha y - \alpha\beta t = 0,$$
$$f_t - \alpha\beta z - \alpha\gamma y - \beta\gamma x + 2\alpha\beta\gamma t = 0.$$

Eliminating α, β, γ, we obtain as the required locus

$$\Sigma \equiv f^2 - yzf_yf_z - zxf_zf_x - xyf_xf_y - tf_xf_yf_z + xyzf_t = 0.$$

If $S \equiv ft + xyz$, we have the identity $S^2 - S_xS_yS_z \equiv \Sigma . t^2$.

This shows that Σ has as nodes the points of contact of tangent planes to S drawn through the lines $x=t=0$, $y=t=0$, $z=t=0$; provided that such points of contact do not lie on $t=0$. Now there are twelve such points of contact, since through any line of S five such tangent planes can be drawn. We have to show that these twelve points of contact form a desmic system. This is seen as follows: it is known that the equation of any cubic surface may be written in the form

$$abt + xyz = 0$$

in 120 ways; if n be the number of cases in which any particular tritangent plane t appears, we have, since the number of tritangent planes is forty-five,

$$n \times 45 = 120 \times 6, \quad \text{hence} \quad n = 16.$$

Hence we may write the equation of the cubic surface in the form

$$abt + (x - \alpha t)(y - \beta t)(z - \gamma t) = 0$$

in sixteen ways; the point of contact of the tangent plane $x - \alpha t$ lies on the line (a, b); and so for the planes $y - \beta t$, $z - \gamma t$. We therefore have twelve points arranged in three groups of four points such that there are sixteen lines each containing *one* point of each group.

Hence the tetrahedra formed by the points of any two groups have four centres of perspective, viz. the points of the third group; the twelve points therefore form a desmic system (Art. 13).

21. The sixteen conics of the surface*.

Along any one of the sixteen lines of the surface three of the coordinates have the same absolute value. Take e.g. the line $y = z = t$; it is easy to see that along this line the tangent plane to the surface is $\lambda y + \mu z + \nu t = 0$; the line is therefore torsal† and the tangent plane meets the surface also in a conic. The surface therefore contains sixteen conics. If the sixteen lines are given, and also the tangent plane along one of them, the surface is determined.

If p is any one of the sixteen lines and π a plane through it, three nodes of the surface lie on p and through each node there pass three of the sixteen lines other than p; thus six of the sixteen lines do not meet p; if $y = z = t$ is the line p, these six lines lie on the quadric

$$x^2 + yz + yt + zt = 0;$$

this quadric is the locus of the conic corresponding to p for different surfaces of the pencil.

The two conics corresponding to two of the sixteen lines which pass through the same node meet in two points, for the line of intersection of their planes passes through the node and meets the surface in two other points lying on these conics. It follows that the four conics which correspond to four lines which intersect each other lie on a quadric; since these lines may intersect in the same node or lie in the same plane, there are twenty-four quadrics each of which meets the surface in four conics.

* Bioche, *Sur les surfaces desmiques du quatrième ordre*, Bull. Soc. math. de France (1909).

† A line at each point of which the tangent plane is the same is said to be *torsal*.

CHAPTER III

QUARTIC SURFACES WITH A DOUBLE CONIC

22. In Chapter I we investigated the quartic surfaces which possess a certain number of isolated singular points; we now consider quartic surfaces which have a double conic.

Any quartic surface with a nodal conic is represented by an equation of the form*

$$\phi^2 = 4w^2\psi,$$

where $\phi = 0$, $\psi = 0$ are quadrics and $w = 0$ is the plane of the double conic. This surface is a variety of syzygetic surface, but the four points given by $\phi = \psi = w = 0$ are here *close-points* on the double conic. At each of them the two tangent planes of the surface coincide with the tangent plane of $\phi = 0$.

This equation may be written

$$(\phi + \lambda w^2)^2 - w^2(\psi + 2\lambda\phi + \lambda^2 w^2) = 0,$$

where λ is arbitrary.

The system of quadrics $\psi + 2\lambda\phi + \lambda^2 w^2$ includes five cones; *every tangent plane of each cone meets the quartic surface in a pair of conics*: for each generator of such a cone is bitangent to the quartic surface, and hence any one of its tangent planes meets the surface in a quartic curve having four nodes, viz. two on the generator of the quadric cone and two where this tangent plane meets the double conic; this quartic curve, therefore, breaks up into two conics, two of whose intersections are collinear with the vertex of the cone.

This may also be seen analytically: for if $B^2 - AC = 0$ is the equation of one of these cones V_1, the equation of the surface is

$$U_1^2 + w^2(AC - B^2) = 0,$$

where $U_1 = \phi + \lambda_1 w^2$.

* Kummer, *Ber. Akad.*, 1863.

It is seen that the equation of the surface involves twenty-one independent constants.

The surface arises as the intersection of two corresponding members of the pencils of quadrics

$$U_1 - wB = \rho w A, \quad U_1 + wB = -\frac{1}{\rho} wC.$$

Each of these quadrics passes through the double conic; the quadrics therefore intersect in another conic, whose plane α is given by

$$\rho^2 A + 2\rho B + C = 0,$$

and this plane is tangent to V_1. The surface is also generated as the intersection of the quadrics

$$U_1 - wB = \frac{1}{\rho} wC, \quad U_1 + wB = -\rho w A,$$

giving the other conic in the plane α.

Hence the tangent planes of V_1 meet the surface in pairs of conics. A similar result arises in connection with each of the cones $V_2 \dots V_5$. Thus the surface contains five sets of ∞^1 pairs of conics. The conics which lie in the tangent planes α of V_1 belong to two classes, viz. those given by

$$\alpha = 0, \quad U_1 = w(B + \rho A),$$

and those given by the equations

$$\alpha = 0, \quad U_1 = -w(B + \rho A).$$

It is clear that two points of intersection of the conics in the plane α lie on the double curve; the other two points lie on the line $\alpha = B + \rho A = 0$, hence they lie on the generator along which α touches V_1.

By considering the conics in two different tangent planes α, β of V_1 it is seen that the conics of the same class do not intersect, and that therefore two conics of different classes intersect twice in points lying on the line (α, β). Among each class of conics which lie in the planes α there are four pairs of lines, arising from those planes α which touch

$$U_1 - w(B + \rho A) = 0 \text{ and } U_1 + w(B + \rho A) = 0$$

respectively. For the condition of tangency gives a quartic for ρ in each case. *Hence the surface contains sixteen lines.*

Each cone $V_2 \dots V_5$ gives rise to eight pairs of lines, but as will be seen, these sets of sixteen lines are the same as the foregoing but differently arranged.

23. Expression of the coordinates in terms of two parameters.

If the cone V_2 has as its equation $B'^2 - A'C' = 0$, we obtain as before two classes of conics on the surface, viz. the intersection of the plane α' or $\sigma^2 A' + 2\sigma B' + C' = 0$ with the quadrics

$$U_2 = w(B' + \sigma A'), \quad U_2 = -w(B' + \sigma A'),$$

where $U_2 = \phi + \lambda_2 w^2$.

There cannot be more than one point common to a conic in the plane α and a conic in the plane α', and therefore *each* conic in α meets *each* conic in α' in *one* point. For instance, a point common to the conic

$$\alpha \equiv \rho^2 A + 2\rho B + C = 0, \quad U_1 = w(B + \rho A),$$

and to the conic

$$\alpha' \equiv \sigma^2 A' + 2\sigma B' + C' = 0, \quad U_2 = w(B' + \sigma A'),$$

must also lie in the plane

$$(\lambda_1 - \lambda_2) w = B - B' + \rho A - \sigma A',$$

and since this plane is not in general coaxal with α and α' there is only one common point.

The coordinates of this point can thus be expressed in terms of two parameters ρ and σ by aid of the last equation and the equations $\alpha = 0$, $\alpha' = 0$. Also the ratios of ρ, σ, and unity are those of three rational functions of the coordinates x_i*.

We thus obtain equations of the form

$$\kappa x_i = F_i(\rho^2\sigma^2, \rho^2\sigma, \rho\sigma^2, \ldots); \qquad (i = 1, 2, 3, 4);$$

the F_i being thus polynomials of the fourth degree in ρ and σ.

These equations in general assign to any given pair of values for ρ and σ one point x_i on the quartic surface, but for such a pair of values of ρ and σ as make the three planes

$$\alpha = 0, \quad \alpha' = 0, \quad (\lambda_1 - \lambda_2) w = B - B' + \rho A - \sigma A',$$

coaxal we have a *line* on the quartic surface, which is the intersection of α and α'. The condition that these planes should be coaxal gives eight sets of values for (ρ, σ); so that if A and A' be a pair of planes which meet in a line of the quartic surface, then $\rho = \infty$, $\sigma = \infty$ gives a line on the surface, and hence in each

* For another proof that a quartic surface with a double conic is rational, see Baker, *Proc. Lond. Math. Soc.* 1912, p. 36.

of the equations $\kappa x_i = F_i$, the coefficient of $\rho^2\sigma^2$ must be zero*, thus giving four equations

$$\kappa x_i = F_i(\rho, \sigma)$$

in which the F_i are cubic functions of ρ and σ.

Making these expressions homogeneous, we obtain

$$\kappa x_i = f_i(\xi_1, \xi_2, \xi_3), \qquad (i = 1, 2, 3, 4),$$

wherein the f_i are of the third degree in the ξ_i, which we may regard as the coordinates of the points of a plane. Since any line meets the quartic surface in four points it follows that the curves of the set $\Sigma a_i f_i = 0$ can have only four variable points of intersection, and hence must have five points in common.

We therefore obtain a (1, 1) *correspondence between the points* x *of the surface and the points* ξ *of the plane, in which plane sections of the surface correspond to, or have as their images, plane cubic curves with five common points*†.

The surface belongs, therefore, to the class of rational surfaces.

24. Mapping of the surface on a plane.

Conversely, starting with the quartic surface which is determined by the equations

$$\rho x_i = f_i(\xi_1, \xi_2, \xi_3),$$

where the curves f_i have five points in common, we can show that it possesses a double conic; for these equations establish a correspondence of such a character that to a plane section there corresponds a plane cubic curve, and since the deficiency of the plane cubic is unity, so also is that of the plane section; the surface, therefore, possesses a double curve of the second order, which must be a conic, since, if it were a pair of non-intersecting straight lines, the surface would be ruled‡.

To each of the five common points of the cubic curves, the *base-points* of the representation, there corresponds a line on the surface; to the points of such a line correspond the points indefinitely near to its corresponding base-point; hence these five lines cannot intersect.

* Otherwise $\rho = \infty$, $\sigma = \infty$ would give a point of the surface.

† This correspondence is taken by Clebsch as the starting point of his investigation of the surface, see Crelle's *Journal*, 1868.

‡ For in this case the line drawn through any point P of the surface to meet these two lines would meet the surface in five points and therefore lie wholly on the surface.

If a curve in the plane passes through the five base-points respectively $\alpha_1 \dots \alpha_5$ times, its image on the surface meets the five lines $a_1 \dots a_5$ times respectively. Let n be the order of any plane curve and N the order of its image on the surface, then since the image of any plane section of the surface is a cubic through the base-points, and since to each point of intersection of the plane curves (not a base-point) there corresponds a point of intersection of their images, we obtain the equation

$$N = 3n - \Sigma\alpha.$$

This equation enables us to determine the curves of different orders which can exist on the surface.

If the curve considered on the surface is a line, $N = 1$, hence

$$1 = 3n - \Sigma\alpha,$$

but each α is either unity or zero, so that the following cases are possible:

$$n = 1, \quad \Sigma\alpha = 2; \qquad n = 2, \quad \Sigma\alpha = 5.$$

In the first case the image of a line on the surface is a line joining two base-points; this gives ten lines on the surface. In the second case the image is the conic through the base-points.

There are, therefore, sixteen and only sixteen lines on the surface.

Each of the sixteen lines is seen to intersect five others, viz. those with which it is paired in the five cones respectively. These five lines do not intersect, as is seen by taking as the image of the first line the conic through the five base-points. It is easy to see, from consideration of the images of the sixteen lines, that they form forty pairs of intersecting lines and forty pairs of twisted quadrilaterals.

25. Conics on the surface.

If in the previous equation we have $N = 2$, we obtain a conic on the surface. Now since none of the α_i can be greater than 2, $n = 4$ would give a plane quartic with five nodes, so that this case must be rejected. Similarly $n = 3$, giving $\Sigma\alpha = 7$, requires that at least two of the α_i should be greater than unity, giving a plane cubic with two nodes; hence the only cases which can arise are:

(i) $n = 2$, one α_i equal to zero, the remainder equal to unity;

(ii) $n = 1$, one α_i equal to unity, the remainder equal to zero.

Hence the image of a conic on the surface is either a conic through four base-points or a line through one base-point.

This gives the ten varieties of conics on the surface previously considered. The two conics in a tangent plane of a cone V correspond to a conic through four base-points and a line through the remaining base-point.

The circumstances of intersection of these various conics are easily deducible from this mode of representation.

The double curve.

The double conic is the only plane section of the surface whose points are not uniquely represented on the plane; its image is a cubic of the family $\Sigma a_i f_i = 0$. The line p whose image is the conic c^2 through the base-points meets the double curve in a point Q which has two images, one P' on c^2 and the other P not on c^2. Every plane section through p meets the surface in a residual cubic through Q; the image of this cubic is a line through P, which forms with c^2 the image of p and the residual cubic. Hence to the cubic in plane sections through p there corresponds the pencil of lines through P.

26. Cubic curves on the surface.

Any plane through one of the sixteen lines meets the surface also in a plane cubic whose image is seen to be *either* a line through P, *or* a conic through three base-points, *or* a cubic through four base-points.

To find the twisted cubics on the surface we again use the equation

$$N = 3n - \Sigma\alpha;$$

since none of the sixteen lines can meet such a cubic in more than two points none of the α_i can be greater than two; hence, if $N = 3$, n is at most equal to four, which would require that four of the α_i should be equal to two, thus giving a plane quartic with four nodes. Hence the only cases are:

(i) $n = 3$, one α_i equal to two, the remaining α_i equal to unity;

(ii) $n = 2$, three α_i equal to unity, the remaining α_i equal to zero;

(iii) $n = 1$, each α_i zero.

This gives sixteen sets of ∞^2 twisted cubics on the surface;

viz. five in the first system, ten in the second and one in the third, each set consisting of ∞^2 cubics.

Two cubics of the same system meet once, cubics of the first and third systems meet three times, cubics of the second and third systems meet twice.

27. Quintic and sextic curves on the surface.

By consideration of the equation $N = 3n - \Sigma\alpha$ we obtain the curves of various orders on the surface. A quadric through a twisted cubic of the first system meets the surface also in a quintic curve whose image (a curve $n = 3$, $\Sigma\alpha = 1 + 1 + 1 + 1$) has deficiency unity; there are ∞^5 such systems of quintics; we denote them by A. The cubics of the second and third systems similarly give rise to systems of quintics, B and C respectively; their images

$$(n = 4,\ \Sigma\alpha = 2 + 2 + 1 + 1 + 1) \text{ and } (n = 5,\ \Sigma\alpha = 2 + 2 + 2 + 2 + 2)$$

are of deficiency unity.

There are also three types of ∞^4 quintics, A', B' and C'*, such that there is one cubic surface which contains a member of A', a conic of the surface and also a member of A; so also for the systems B' and C'. A system D of ∞^4 quintics exists such that one cubic surface can be determined to contain a quintic of this system and also two non-intersecting lines of the surface.

One cubic surface exists which contains any sextic curve on the quartic surface; it will meet the surface in another sextic. We obtain three varieties of sextics†, viz.:

∞^5 sextics lying in pairs on the cubic surface whose images have deficiency zero,

∞^6 sextics lying in pairs on the cubic surface whose images have deficiency unity,

∞^7 sextics lying in pairs on the cubic surface whose images have deficiency two.

* Their images are respectively given by

$$n=4,\ \Sigma\alpha=3+1+1+1+1\,;\quad n=3,\ \Sigma\alpha=2+1+1\,;\quad n=2,\ \Sigma\alpha=1.$$

† Their images are

$$n=2,\ \Sigma\alpha=0\,;\quad n=3,\ \Sigma\alpha=1+1+1\,;\quad n=4,\ \Sigma\alpha=2+1+1+1+1.$$

28. Quartic curves on the surface.

Assuming the existence of a twisted quartic curve on the surface, an infinite number of quadrics will pass through it if it is of the first species and one quadric if it is of the second species; thus at least one other twisted quartic exists on the surface. Each quartic on the surface gives rise to an image and since the order of the image of the complete curve of intersection of the surface and any quadric is six*, the sum of the orders of the images of the two quartics is also six; therefore the image of a twisted quartic is of the order 2, 3 or 4.

Since no α_i can be greater than two, the equation $4 = 3n - \Sigma\alpha$ allows of the following solutions:

(1) $n = 2$, two α_i equal to unity, the rest equal to zero;

(2) $n = 4$, three α_i equal to two, two α_i equal to unity;

(3) $n = 3$, three α_i equal to unity, one α_i equal to two, one α_i equal to zero;

(4) $n = 3$, each α_i equal to unity.

This gives rise to forty-one sets of twisted quartics; viz. from (1) and (2) arise ten sets, (3) gives twenty, (4) gives 1. The quartics in (1) and (2) lie in pairs on a quadric, those in (3) lie in pairs on a quadric. The quartics (4) arise as the intersections with the surface of the quadrics through the double conic.

Each class consists of ∞^3 members except class (4) which contains ∞^4.

In each of the first three classes the corresponding quartics are of the second species; for through three points of intersection (not base-points) of any two curves of the system $\sum_1^4 k_i f_i = 0$ one curve of each of the first three systems can be drawn; hence the corresponding quartic curves possess a trisecant and therefore belong to the second species. But any cubic of the fourth class which passes through three points of intersection of two curves of the system $\sum_1^4 k_i f_i = 0$ must itself belong to this system and therefore correspond to a plane curve; hence the fourth class does not possess trisecants.

From consideration of the curves whose images belong to either of the cases (1), (2), or (3) it is clear that two curves belonging to the same set

* Since if n and n' are the orders of the two images, $8 = 3(n + n') - 10$.

intersect in two points, two curves on the same quadric in six points. Two quadri-quartics on the same quadric intersect in eight points, but since their images intersect in only four points, it is clear that they meet on the double curve, but on different sheets of the surface, four times.

29. Class of the surface.

The class of the surface is *twelve*: for if a plane through a given line touches the surface its curve of intersection with the surface has a node at the point of contact; hence the corresponding cubic curve has a node, therefore the number of tangent planes of the surface through a given line is equal to the number of nodal cubics of the family

$$\Sigma\lambda_i f_i = 0$$

subject to the two conditions $\Sigma\lambda_i a_i = 0$, $\Sigma\lambda_i b_i = 0$; that is to the number of nodal cubics of the pencil $S + \rho S' = 0$, where $S = 0$, $S' = 0$ are two members of the family. Now if $S + \rho S' = 0$ has a node its discriminant vanishes, and this discriminant is of degree twelve in ρ; hence the required class of the surface is twelve.

30. The sixteen lines of the surface.

It will now be shown that the relationship between the sixteen lines of the surface, as regards mutual intersection, is identical with that which exists between sixteen lines selected in a certain manner* of the general cubic surface.

For the equation of the general cubic being

$$\begin{vmatrix} a & b & c \\ a' & b' & c' \\ a'' & b'' & c'' \end{vmatrix} = 0,$$

where $a, b, \ldots$ are linear in the variables, is equivalent to the following:

$$\begin{aligned} \xi_1 a + \xi_2 b + \xi_3 c &= 0, \\ \xi_1 a' + \xi_2 b' + \xi_3 c' &= 0, \\ \xi_1 a'' + \xi_2 b'' + \xi_3 c'' &= 0. \end{aligned}$$

The last equations lead, on solving for $x_1 \ldots x_4$, to the equations

$$\rho x_1 = f_1(\xi_1, \xi_2, \xi_3), \quad \rho x_2 = f_2(\xi_1, \xi_2, \xi_3),$$
$$\rho x_3 = f_3(\xi_1, \xi_2, \xi_3), \quad \rho x_4 = f_4(\xi_1, \xi_2, \xi_3);$$

* See Geiser, *Ueber die Flächen vierten Grades welche eine Doppelcurve zweiten Grades haben*, Crelle, LXX.

in which the f_i are of the third degree, and the curves $f_i = 0$ have six points in common (this follows from the fact that any line meets the cubic surface in three points, and therefore any two members of the family $\Sigma k_i f_i = 0$ have three variable points of intersection).

We thus establish a (1, 1) correspondence between the points of the cubic surface and those of the plane, and since such a correspondence is already established between the plane and the quartic surface the points of the cubic and quartic surfaces are themselves so connected.

In the transformation expressed by the last equations there are thus six base-points $P_1 \ldots P_6$, the first five of which we may suppose to be base-points in the transformation connected with the quartic surface: denoting the surfaces by C_3 and C_4 respectively, to the points $P_1 \ldots P_5$ there correspond in the two surfaces the lines $\pi_1 \ldots \pi_5$ and $p_1 \ldots p_5$ respectively; since the equation $N = 3n - \Sigma\alpha$ holds also for the cubic surface we deduce as in the case of C_4 that to the joins of $P_1 \ldots P_5$ there respectively correspond the lines $\lambda_{12} \ldots \lambda_{45}$ in C_3, and $l_{12} \ldots l_{45}$ in C_4; finally to the conic through $P_1 \ldots P_5$ there correspond the lines Λ_6 and L_6.

Hence the sixteen lines on the two surfaces are connected as follows: to

$$p_1 \ldots p_5;\ l_{12} \ldots l_{45};\ L_6$$

there correspond

$$\pi_1 \ldots \pi_5;\ \lambda_{12} \ldots \lambda_{45};\ \Lambda_6.$$

Thus the relationship of the sixteen lines on C_3 as regards intersection is the same as that of the corresponding lines on C_4, being deduced in both cases from the relationship of the base-points and lines in the plane.

Now the sixteen lines of C_3 are obtained by omitting from its twenty-seven lines, π_6 and the ten lines which meet π_6, hence the sixteen lines of the quartic surface are obtained by omitting from the twenty-seven lines of a general cubic surface any one of these lines and the ten lines which intersect it.

31. Determination of the surface by aid of two quadrics and a given point.

The surface may be obtained by aid of any two given quadrics and a given point, a_i. For let P, P' be two points x_i, x_i' collinear with α_i and conjugate for a given quadric H, let K be the fourth harmonic point for α_i, P, P'; then if K is the point y_i we have

$$x_i = \sigma\alpha_i + \tau y_i, \quad x_i' = \sigma\alpha_i - \tau y_i;$$

$$x_i' = \frac{2H}{\Delta H}\alpha_i - x_i, \quad \text{where } \Delta H = \sum_1^4 \alpha_i \frac{\partial H}{\partial x_i}.$$

These equations lead to

$$\rho y_i = x_i \Delta H - \alpha_i H, \qquad (i = 1, 2, 3, 4).$$

If now K describes the quadric $Q_y = 0$, we have

$$Q_x (\Delta H)^2 - H_x \Delta H \Delta Q + H_x^2 Q_\alpha = 0,$$

which may be written in the form

$$(2HQ_\alpha - \Delta H \Delta Q)^2 = (\Delta H)^2 \{(\Delta Q)^2 - 4Q_x Q_\alpha\}.$$

This represents a general quartic surface with a double conic whose plane is the polar plane of α_i for H; one of the cones of Kummer is the tangent cone to Q whose vertex is α_i.

Writing this equation, as before, in the form

$$\{2HQ_\alpha - \Delta H \Delta Q - \lambda (\Delta H)^2\}^2$$
$$= (\Delta H)^2 \{(\Delta Q)^2 - 4Q_x Q_\alpha - 2\lambda (2HQ_\alpha - \Delta H \Delta Q) + \lambda^2 (\Delta H)^2\},$$

then if $\lambda H + Q = W$, we obtain finally an equation of the form

$$U^2 = (\Delta H)^2 \{(\Delta W)^2 - 4Q_\alpha W\}.$$

If W is one of the four cones through the intersection of H and Q, the last factor is a quadric touching W along two lines, i.e. it is a cone with the same vertex: hence, *the vertices of the four remaining cones of Kummer are those of the tetrahedron self-polar for Q and H**.

The following result may be deduced: *the five lines joining any point P on the double conic to the vertices of the cones of Kummer and the tangent to the double conic at P, lie on a quadric cone.*

If Q and H touch, the point of contact is a node of the quartic surface, for the equation of the latter being

$$\Delta H (Q \Delta H - H \Delta Q) + H^2 Q_\alpha = 0,$$

if Q and H touch, their point of contact is a node of the cubic surface $Q\Delta H - H\Delta Q = 0$, and hence a node on the quartic surface†.

* Bobek, *Ueber Flächen vierter Ord. mit einem Doppelkegelschnitte*, Sitzb. d. K. Akad. Wien, 1884.

† It is easy to see that this cubic surface is the locus of points P, P' which are collinear with α_i and conjugate for both Q and H.

From the foregoing method it is seen that both P and P' lie on the quartic surface.

When K describes a line p of one regulus belonging to Q, then P and P' lie on a conic for which p is the polar line of α_i; when K describes the line p' lying in the plane (p, α) and belonging to the other regulus of Q, the points P, P' describe another conic in this plane. We thus obtain the two sets of conics lying in the various tangent planes to Q which pass through α_i. These planes envelop the tangent cone to Q whose vertex is α_i.

Coincidence of the points P, P' occurs at the points in which p meets the conic; both these points lie on Q and on H. If p touches H it will follow that the intersections of the conic and its polar line for α come into coincidence; hence the conic must in this case become a pair of lines. Since *four* of the lines of any regulus touch any quadric, we obtain the sixteen lines of the surface.

Three pairs of lines belonging to one of the two classes associated with a cone of Kummer determine the surface; for every conic of the other class meets each of the six lines (Art. 25), hence the ∞^1 planes through the intersection of the planes of the three pairs of lines such that each plane meets the lines in six points on a conic will envelop the corresponding cone of Kummer, and the surface is determined.

Assuming the six lines to have general positions, the number of constants involved is twenty-one* (Art. 22).

The quadric Q meets the surface in two quadri-quartic curves, one of them $Q = H = 0$ is the curve of contact of the residual tangent cone drawn to the surface from the vertex of the cone of Kummer.

32. Perspective relation with a general cubic surface.

It has been seen (Art. 23) that a (1, 1) correspondence can be established between a quartic surface with a nodal conic and a plane.

Two methods† have been given of establishing a (1, 1) perspective correspondence between the points of a general cubic surface and the quartic surface. Two points x, x' are collinear

* Weiler, *Ueber Flächen vierter Ord. mit Doppel- und mit Cuspidal-Kegelschnitten*, Schlömilch Zeitsch. xxx.

† Geiser, *l.c.*; Cremona.

with A_4* and conjugate for the quadric $\sum_1^4 x_i^2 = 0$ if the coordinates are connected by the equations

$$\rho x_1 = x_1' x_4', \quad \rho x_2 = x_2' x_4', \quad \rho x_3 = x_3' x_4',$$
$$\rho x_4 = -(x_1'^2 + x_2'^2 + x_3'^2).$$

Taking any cubic surface through the curve

$$x_4 = x_1^2 + x_2^2 + x_3^2 = 0$$

which does not pass through A_4, and whose equation is therefore of the form

$$x_4 U + (x_1^2 + x_2^2 + x_3^2) L = 0,$$

where $U = 0$ is any quadric and $L = 0$ any plane; on transformation $x_1^2 + x_2^2 + x_3^2$ becomes $x_4'^2 (x_1'^2 + x_2'^2 + x_3'^2)$, $L = 0$ becomes a quadric through the conic

$$x_4' = x_1'^2 + x_2'^2 + x_3'^2 = 0,$$

and $U = 0$ becomes a quartic surface having this conic as double curve; omitting the factor $x_1'^2 + x_2'^2 + x_3'^2$ we therefore obtain a quartic surface with the double conic

$$x_4' = x_1'^2 + x_2'^2 + x_3'^2 = 0.$$

The second transformation is the following: the points x, x' are connected by the equations

$$x_1 : x_2 : x_3 : x_4 = S_1(x') : S_2(x') : S_3(x') : S_4(x'),$$

where the quadrics $S_i(x') = 0$ all pass through a given conic and touch each other at a given point on this conic.

By linear combination it is seen that this is equivalent to the transformation

$$x_1 : x_2 : x_3 : x_4 = x_1'^2 : x_1' x_2' : x_1' x_3' : x_2' x_4' - x_3'^2;$$

from which we deduce that

$$x_1' : x_2' : x_3' : x_4' = x_1 x_2 : x_2^2 : x_2 x_3 : x_1 x_4 + x_3^2.$$

In this case the centre of projection, the point A_4, lies on the conic. This transformation is of the (1, 1) character, the exceptions being that to the point A_4 for x there corresponds the plane x_1', to the point A_4 for x' there corresponds the plane x_2, and to any point on x_2, which is not also on the conic $x_2 = x_1 x_4 + x_3^2 = 0$, there corresponds the *same point* A_4 in the field of x'.

To the planes in the field of x there correspond quadrics which pass through the fixed conic in the field of x', but to a plane

* Here, and elsewhere, the vertices of the tetrahedron of reference will be denoted by $A_1 \dots A_4$.

through A_4 in one field there corresponds the same plane in the other field.

Having given a general cubic surface f in the field of x which contains the conic $x_2 = x_1x_4 + x_3^2 = 0$, it is seen as before that f is projected into a general quartic surface F with the nodal conic

$$x_1' = x_2'x_4' - x_3'^2 = 0.$$

The section of f by x_2 consists of the given conic together with a line a to which the point A_4 corresponds in the field of x'. Let the ten lines of f which meet a be denoted by

$$(b_1, c_1), \quad (b_2, c_2), \quad (b_3, c_3), \quad (b_4, c_4), \quad (b_5, c_5).$$

Consider the plane through b_1 and c_1; it corresponds to a quadric through the double conic in the field of x' which therefore meets F in a twisted quartic which accordingly must consist of two conics, each passing through A_4.

Hence to the sections of f by the planes (A_4, b_1), (A_4, c_1) correspond four conics through A_4; hence we have twenty conics through A_4. But if d is a line of f which meets the conic

$$x_2 = x_1x_4 + x_3^2 = 0,$$

then to the section of f by the plane (A_4, d) there corresponds the section of F by the same plane which therefore meets F in a line and a cubic, since all the sections through A_4 consisting of two conics are given by the twenty preceding conics.

Since there are sixteen lines such as d, it follows that the sixteen lines of the quartic surface are thus projectively derived from the sixteen lines of f.

The ten sets of conics on F are the images of the conics of f in the ten pencils of planes whose axes are the lines $b_1 \ldots c_5$.

33. Projective formation of the surface.

Bobek* has developed a method of treatment of the surface depending upon its formation from two pencils of quadrics projectively related. If

$$\lambda ab + ad + U = 0, \quad \mu ac + ae - U = 0,$$

represent two pencils of quadrics each passing through a fixed conic (a, U) and through two other fixed conics respectively, then any two members of these pencils intersect in another conic whose plane is

$$\lambda b + \mu c + d + e = 0;$$

this plane passes through the fixed point $b = c = d + e = 0$.

If the pencils are connected by a given lineo-linear relation between λ and μ, it is clear that the locus of intersection of corresponding members of the two

* *Loc. cit.*

pencils is a quartic surface with (a, U) as double conic. Moreover the above planes will in this case touch a cone whose vertex is the aforesaid fixed point, and the conics in these planes will form the conics on the quartic surface. Taking the equation of the quartic surface in the form

$$a^2(bc+d^2)=U^2,$$

the foregoing two pencils are

$$\lambda ab = U - ad, \quad \mu ac = U + ad,$$

with the relation $\lambda\mu = 1$.

Any line through the point $b=c=d=0$, the vertex of the cone of Kummer selected, meets the first pencil of quadrics in pairs of points in involution; if $U \equiv aK + V$, the double points of this involution are obtained from the equation

$$V = a^2,$$

which is also the locus of double points of the involution determined on lines through the vertex by quadrics of the second pencil. It is the surface H previously given.

The surface Q appears as the locus of the line of intersection of the polar planes of a pair of corresponding quadrics for the vertex of the given cone.

34. Connection of properties of the surface with those of plane quartics.

Zeuthen* has investigated the plane quartic which is the section of the tangent cone to the surface from any point of the double conic, and showed its relationship to the surface.

Taking the equation of the surface as being $U^2 + z^2W = 0$ we may assume any point P on the double conic as that through which the coordinate planes x, y, z pass, and take the polar plane of P for W as the fourth coordinate plane $t = 0$; let the tangent plane to U at P be the plane $y = 0$.

The equation of the surface then becomes

$$a(\psi + yt)^2 + bz^2(\phi + t^2) = 0.$$

Writing this in the form

$$t^2(bz^2 + ay^2) + 2ta\psi y + a\psi^2 + b\phi z^2 = 0,$$

the tangent cone from P to the surface has as its equation

$$\phi(ay^2 + bz^2) + a\psi^2 = 0.$$

This is a general quartic cone, having the planes $ay^2 + bz^2 = 0$, the tangent planes to the surface at P, as bitangent planes.

Hence any plane quartic may be regarded as the "projection" of a nodal quartic surface from any point on the double conic.

* *Sulle superficie di quarto ordine con conica doppia*, Ann. di Mat. II. XIV. (1887).

Now having given any pair of bitangents $ay^2 + bz^2 = 0$ of a quartic curve, the equation of the curve may be written in the form

$$(ay^2 + bz^2)\,\alpha\beta = V^2$$

in five ways*, giving a *group* of six bitangents, and in consequence the equation of the surface may be written in the form

$$a(\psi + yt)^2 + bz^2(\alpha\beta + t^2) = 0,$$

giving five pairs of planes α, β through P which meet the surface in a pair of conics. They are the tangent planes from P to the five cones of Kummer, and bitangent planes of the tangent cone of vertex P.

There remain sixteen of the twenty-eight bitangents of the quartic curve, giving rise to sixteen bitangent planes of the cone; each plane meets the surface in a quartic curve having four double points (of which one is at P, and another is the second intersection of the plane with the nodal conic). This curve will consist of a line and a cubic curve having a node at P; thus the existence of the sixteen lines of the surface becomes manifest.

If again three coordinate planes be taken as passing through the vertex of one of the cones of Kummer, the plane x being that which does not pass through the preceding point P, the equation of the surface may be taken to be

$$(\phi + xL)^2 = z^2(yt + x^2).$$

The equation of the preceding tangent cone of vertex P is then

$$yt(z^2 - L^2) = \phi^2 \;\ldots\ldots\ldots\ldots\ldots\ldots\ldots (1).$$

Now the cone

$$\rho^2 y(L + z) - 2\rho\phi - t(L - z) = 0 \;\ldots\ldots\ldots\ldots (2)$$

touches the cone (1) along four lines, and the plane

$$\rho^2 y + 2\rho x - t = 0$$

meets the cone (2) in a conic which is seen, by elimination of ρ, to lie on the quartic surface.

Regarding the equations (1) and (2) as representing curves, it is seen that *the four-point-contact conics* (2) *are the projections from P of a system of conics of the surface.*

The other system connected with this cone of Kummer gives rise to the four-point-contact conics

$$\rho^2 y(L - z) - 2\rho\phi - t(L + z) = 0.$$

Theorems relating to the four-point-contact conics of a quartic curve are thus connected with theorems concerning this quartic

* See Salmon, *Higher Plane Curves.*

surface; e.g. take the theorem: *the eight points of contact with the quartic of any two conics of such a system lie on one conic**.

We obtain the theorem for the quartic surface: *the two pairs of principal tangents at a point P of the double conic and the points of contact with the surface of the two planes through P which touch the same cone of Kummer, lie on a quadric cone†*.

Again in Art. 31 it was seen that the intersection of the tangent planes at P to the surface and the vertices of the five cones of Kummer lie on a quadric cone whose vertex is P; hence we derive the result for quartic curves that *the six intersections of pairs of bitangents of a group lie on a conic.*

It has been seen that the group of six pairs of bitangents determined by the tangent planes to the surface at P gives four-point-contact conics which are the projections from P of the conics of the surface. It will now be shown that the other four-point-contact conics are projections of cubics on the surface.

Refer the surface to coordinate planes consisting of the plane of the double conic and three tangent planes of a cone of Kummer of which one, x, contains two lines of the quartic surface; the equation of the surface is then of the form

$$\{AB + z(y-t) + xL\}^2 = z^2\{x^2 + y^2 + t^2 - 2xy - 2xt - 2yt\}.$$

The equation of the quartic tangent cone whose vertex is P is then

$$\{y(z+L) + t(z-L) + AB\}^2 = 4ABt(z-L);$$

of which a four-line-contact cone is

$$\rho^2 At + \rho\{y(z+L) + t(z-L) + AB\} + B(z-L) = 0.$$

This meets the cubic surface

$$2Azt\rho = (L-z)\{x(L+z) + AB\}$$

in the line $\qquad L + z = 0, \quad \rho t + B = 0,$

which passes through P, and also in a quintic curve having a triple point at P and which lies on the quartic surface.

Hence the preceding quadric cone also meets the quartic surface in a cubic curve passing through P.

* For the quartic $(z^2 - L^2)\psi = \phi^2$ may be written

$$(z^2 - L^2)(\lambda^2\psi + 2\lambda\phi + z^2 - L^2) = (\lambda\phi + z^2 - L^2)^2.$$

The points of contact are given as the intersections of the conics

$$\lambda^2\psi + 2\lambda\phi + z^2 - L^2 = 0, \quad \lambda\phi + z^2 - L^2 = 0;$$

moreover the conic $\lambda\mu\psi + (\lambda+\mu)\phi + z^2 - L^2 = 0$ passes through them and also through the four points similarly obtained on replacing λ by μ.

† The principal tangents being $z^2 - L^2 = \phi = 0$, and taking $\psi \equiv yt$ and $\lambda = \infty$, $\lambda = 0$ successively.

35. Segre's method of projection in four-dimensional space.

Segre has shown* that if $F=0$, $\Phi=0$ are two quadratic manifolds or *varieties* in flat space of four dimensions S_4, the projection upon any hyperplane S_3 of their intersection Γ, is a quartic surface with a double conic. For if A, or x', be any point of S_4 the substitution of $x_i'+\rho x_i$ for x_i in $F=0$ gives the two intersections of the line (x, x') with F. The elimination of ρ between the equations

$$F_{x'}+\rho DF+\rho^2F_x=0,$$
$$\Phi_{x'}+\rho D\Phi+\rho^2\Phi_x=0,$$

gives the "cone" joining A to the points of Γ. The intersection of this cone with the hyperplane S_3 or $\sum_1^5 a_ix_i=0$, gives a surface in S_3 represented by the equation

$$(F\Phi'-F'\Phi)^2-(\Phi DF-FD\Phi)(F'D\Phi-\Phi'DF)=\sum_1^5 a_ix_i=0.$$

Taking F to be f, that member of the pencil (F, Φ) which passes through A, since $f(x')=0$, we may write as the equation of the projected surface

$$f^2\phi'-Df(fD\phi-\phi Df)=\sum_1^5 a_ix_i=0;$$

that is

$$(2f\phi'-Df.D\phi)^2-(Df)^2\{(D\phi)^2-4\phi\phi'\}=\sum_1^5 a_ix_i=0\dagger.$$

This is a quartic surface with the nodal conic

$$Df=f=\sum_1^5 a_ix_i=0.$$

It is seen that the double conic is obtained as the intersection of f and Df, since the only cases in which the line joining A to any point x meets Γ in two points are when the foregoing quadratics in ρ become identical, we then have

$$F_{x'}\Phi_x-\Phi_{x'}F_x=0,\quad F'D\Phi-\Phi'DF=0.$$

These equations represent respectively the variety f through A and its *tangent hyperplane* Df. Their intersection gives a quadric cone in three dimensions which meets any variety of the pencil

* *Surfaces du quatrième ordre à conique double*, Math. Ann. XXIV. For many details the reader is referred to this important memoir.

† Compare with Art. 31.

(F, Φ), and therefore Γ, in a twisted quadri-quartic k^4. This quartic k^4 is projected upon S_3 as a conic; any generator of the cone meets k^4 in two points P, Q; the tangent planes to Γ at P and Q are projected into the tangent planes of the quartic surface at a point of this conic.

Among the generators of the cone (f, Df) there are in general four which touch k^4 (viz. at the points where the plane $Df=0$, $D\phi=0$ meets Γ). It follows that there are four *pinch-points* on the double conic.

There are in general five cones in the pencil (F, Φ). For if F and Φ are not specially related to each other we may take

$$F \equiv \sum_1^5 x_i^2, \quad \Phi = \sum^5 a_i x_i^2;$$

the pencil therefore contains the five cones

$$(a_2 - a_1)\, x_2^2 + (a_3 - a_1)\, x_3^2 + (a_4 - a_1)\, x_4^2 + (a_5 - a_1)\, x_5^2 = 0, \text{ etc.}$$

If f is a cone, i.e. if A lies on one of the cones of the pencil (F, Φ), we have two double lines instead of a double conic. For the hyperplane Df meets f in two planes*, the intersection of these planes with Γ will consist of two conics having two common points lying on a line through x'. These conics are projected from x' into two intersecting double lines of the quartic surface.

Any one of the five cones of the pencil (F, Φ) may be represented by an equation of the form $\sum_1^4 a_i x_i^2 = 0$, whence by comparison with the general three-dimensional quadric it is seen that this cone possesses two sets of generating planes, each generating plane of one set meets each generating plane of the other set in a line, the two planes therefore lie in the same hyperplane, while two generating planes of the same set intersect only at the vertex of the cone†.

* For we may take the cone f to be $\sum_1^4 a_i x_i^2 = 0$, the tangent hyperplane to this cone at a point x' is $\sum_1^4 a_i x_i x_i' = 0$; interpreting these equations to represent a quadric and its tangent plane at x', since the plane meets the quadric in *two lines*, the hyperplane Df will meet f in *two planes* whose intersection contains x'.

† It will be seen hereafter (see Art. 49), that the pencil (F, Φ) may contain, in certain cases, a cone of the *second species*, i.e. a cone whose equation contains only three variables, e.g. x_1, x_2, x_3; in this case the generating planes consist of a simply infinite set of planes passing through the line $x_1 = x_2 = x_3 = 0$.

Each generating plane of a cone meets Γ in a conic; conversely each conic, c^2, of Γ lies in a generating plane of a cone of the pencil (F, Φ); for the variety of a pencil which passes through any point P in the plane of c^2 and not upon c^2, must contain the plane entirely, and a variety which contains a plane is necessarily a cone*.

Hence Γ contains ∞^1 conics belonging to five sets, each set containing two classes (corresponding to the two systems of generating planes of a cone).

The hyperplane through any generating plane α of a cone and A meets the cone in another generating plane α'; for taking the cone as $x_1x_2 - x_3x_4 = 0$, and the generating plane as

$$x_1 - \mu x_3 = \mu x_2 - x_4 = 0,$$

the hyperplane is

$$(x_1 - \mu x_3)(\mu x_2' - x_4') - (\mu x_2 - x_4)(x_1' - \mu x_3') = 0 \quad \text{......(1).}$$

By comparison with the three-dimensional quadric it follows that this hyperplane also contains another generating plane α' belonging to the other system. Since α and α' belong to the same hyperplane (through A), it follows that they are projected from A into the same plane β of S_3, and in β there lie two conics of the quartic surface. The envelope of β is seen from (1) to be a quadric cone whose vertex is the projection of the vertex (00001). Thus we regain the pair of conics in each tangent plane of a cone of Kummer; the points of intersection of such a pair lying on the generator of the cone along which the plane touches the cone.

Other leading properties of the quartic surface considered are readily obtained by the method of Segre. We obtain the sixteen lines of the surface as follows:

The surface Γ is determined by the equations

$$\sum_2^5 (a_i - a_1) x_i^2 = 0, \quad \sum_1^5 x_i^2 = 0;$$

by a change of the coordinate system these equations may be replaced by

$$X_1X_2 - X_3X_4 = 0 \quad \text{....................}(a)$$

$$x_1^2 + (a \between X_1, X_2, X_3, X_4)^2 = 0 \quad \text{..............}(b).$$

* Since its equation is expressible in the form $x_1A + x_2B = 0$, if $x_1 = x_2 = 0$ is the given plane.

Every plane $X_1 = \lambda X_3$, $X_4 = \lambda X_2$ (a generating plane of (a)), will meet (b) in a conic, which reduces to two lines, if

$$(a)(X_1, X_2, X_3, X_4)^2$$

is reduced to a perfect square; this leads to a biquadratic in λ.

Hence four generating planes of this system meet Γ in two lines, and similarly four generating planes of the other system meet Γ in two lines; this gives sixteen lines on Γ, and therefore, by projection, on the nodal quartic surface.

It follows also that eight tangent planes of each cone of Kummer contain a pair of these lines.

Each of the sixteen lines p on Γ lies on each of the five cones of the system; the plane through p and a vertex of one of these cones is a generating plane of that cone and therefore meets Γ in another line, hence each of the sixteen lines is met by five others.

Cubics and Quartics on the surface.

Any hyperplane through one of the sixteen lines meets Γ in a cubic curve, and since there are ∞^2 hyperplanes through any line we thus obtain sixteen sets of ∞^2 cubic curves on the surface.

In S_4 there are ∞^4 hyperplanes and each of them meets Γ in a quadri-quartic, any two of these quadri-quartics intersect in four points, lying in the plane common to the two hyperplanes; through these four points there pass ∞^1 quadri-quartics determined by the pencil of hyperplanes through the plane of the four points.

Since four non-coplanar points determine one hyperplane, it follows that one quadri-quartic of the surface passes through any four non-coplanar points of a nodal quartic surface.

Any hyperplane $\Sigma = 0$ cuts the quadri-quartic k^4 whose projection is the double conic, in four points lying in the plane of intersection of this hyperplane with Df, the tangent hyperplane at A.

Let $Q_1 \dots Q_4$ be these four points and α their plane, and $Q_1' \dots Q_4'$ the points in which AQ, etc. again meet k^4. Let β be the polar plane of A for the system of quadrics through k^4 and p the line $(\alpha\beta)$; then the planes (pA), α, β, (pQ_1') are harmonic and the plane (pQ_1') must pass through $Q_2' \dots Q_4'$; i.e. the points $Q_1' \dots Q_4'$ are coplanar. Hence we have ∞^1 quadri-quartics, arising from the hyperplanes $\Sigma + \lambda Df = 0$, through $Q_1 \dots Q_4$, and ∞^1 quadri-quartics through $Q_1' \dots Q_4'$. It follows on projection that *each of the* ∞^4

quadri-quartics of the projected surface cuts the double conic in four points of which three determine the fourth; and through four such points there pass ∞^1 quadri-quartics on one sheet of the surface and ∞^1 quadri-quartics on the other sheet of the surface.

Among the ∞^4 hyperplanes of S_4, ∞^3 pass through A; these hyperplanes meet Γ in quadri-quartics which are projected into *plane sections* of the projected surface.

Quadrics inscribed in the surface.

Let $F=0$ be any variety of the pencil (F, Φ); its intersection with the polar hyperplane of A for F is given by $DF=0$, $F=0$, and is a quadric Λ; the intersection of Γ with Λ is a quadri-quartic c^4. Let X be any point of c^4; the tangent plane to Γ at X is given by the equations

$$\Sigma X_i \frac{\partial F}{\partial x_i} = 0, \quad \Sigma X_i \frac{\partial \Phi}{\partial x_i} = 0,$$

and the tangent plane to Λ at X is given by

$$\Sigma X_i \frac{\partial F}{\partial x_i} = 0, \quad DF = 0;$$

each of these tangent planes lies in the hyperplane $\Sigma X_i \frac{\partial F}{\partial x_i} = 0$, which passes through A since X is a point on $DF=0$. Hence the tangent planes to Γ and to Λ at X lie in the same hyperplane through A, they are therefore projected into the same plane of S_3.

Thus the projection of Λ touches the projection of Γ along a quadri-quartic, the projection of c^4.

Now F is *any* member of the pencil (F, Φ), hence ∞^1 *quadrics touch the quartic surface along quadri-quartic curves.*

36. Fundamental inversions.

As in the case of the quadric in three dimensions where the points of contact of the tangent lines to a quadric ϕ which pass through a point lie on a plane, so the points of contact of tangents passing through a point of S_4 lie on a hyperplane, the polar hyperplane of the point. If C is the vertex of a cone of the pencil (F, Φ), the polar hyperplane of C is the same for each member of the pencil, e.g. if the system is determined by the two equations

$$\sum_1^5 x_i^2 = 0, \quad \sum_1^5 a_i x_i^2 = 0,$$

the polar hyperplane of the point (10000) is $x_1 = 0$, and so on.

Let α be the polar hyperplane of C, A the centre of projection, and f the member of the pencil which passes through A. Then any plane through the line (C, A) meets f in a conic passing through A: this conic is met by $\alpha = 0$ in two points B, B' whose join is the polar line of C for this conic; so that if any line through C meets the conic in two points Q, Q' then, by elementary geometry, $\{A, BB'QQ'\} = -1$.

Now there are two generators of the cone whose vertex is C which lie in the plane of this conic; each of these generators meets the conic in two points of Γ, since the points lie both on $f = 0$ and on the cone, and the conic is projected from A into a line of S_3, hence denoting by σ the quadric which is the projection of the quadric $\alpha = 0$, $f = 0$, and the projection of C by C' (which is the vertex of a cone of Kummer), it follows that any line through C' meets σ in two points B_1, B_1' and the projected quartic surface in two pairs of points Q_1, Q_1'; R_1, R_1' such that both Q_1, Q_1' and R_1, R_1' are harmonic with regard to B_1, B_1'.

Hence C' is said to be a *centre of self-inversion* of the projected quartic surface*.

37. Plane representation of the surface.

To represent Γ, and therefore the projection of Γ, upon a plane, we take the ∞^2 planes through a line p of Γ, any one of these planes meets any two varieties of the pencil (F, Φ) in p and two other lines respectively, the intersection, Q, of these latter lines lies on Γ; hence the plane through p meets Γ in one other point, viz. Q. Moreover it meets any given plane K in one point Q', thus there arises a (1, 1) correspondence between the points of Γ and K.

The five lines of Γ which meet p have as images the five base-points; if q be one of the ten lines which do not meet p, the hyperplane through p and q meets K in a line, and since this hyperplane meets Γ in two non-intersecting lines p and q, it must also meet it in two other lines which meet both p and q. Hence the image of q is a line passing through two base-points.

If $x_1 = 0$, $x_2 = 0$ are tangent hyperplanes to F at any two points of p, and $x_3 = 0$, $x_4 = 0$ the tangent hyperplanes to Φ at these points, the tangent plane to Γ at any point of p is represented by

$$x_1 + \lambda x_2 = 0, \quad x_3 + \lambda x_4 = 0.$$

* The quadric σ is the quadric H of Art. 31.

As λ varies, the intersection of this plane with the given plane K is clearly a conic; hence the points of Γ contiguous to p are represented by the points of a conic which passes through the five base-points, i.e. *the image of p* is this conic.

Again the ∞^2 hyperplanes through p meet Γ in ∞^2 cubics and K in ∞^2 lines, i.e. the lines of K are the images of ∞^2 cubics of Γ.

Any hyperplane meets Γ in a quadri-quartic and also meets each of the five lines which meet p, moreover it meets any cubic of Γ in three points, hence the image of the section of Γ by this hyperplane is such that it is met by any line of K in three points; it is therefore a cubic which passes through the five base-points.

The ∞^3 hyperplanes through A which give rise to the plane sections of the projected surface (Art. 35) meet Γ in quadri-quartics such that through any three points of Γ there passes one such quadri-quartic, hence among the ∞^4 cubics of K through the base-points there are ∞^3 cubics forming a *net* or linear set; these are the images of the plane sections of the projected surface.

CHAPTER IV

QUARTIC SURFACES WITH A NODAL CONIC AND ALSO ISOLATED NODES

38. A quartic surface with a nodal conic may have in addition one or more isolated nodes; such a node is the vertex of a cone of Kummer, for taking the node as a vertex of the tetrahedron of reference, the equation of the surface is

$$x_4^2 A + 2x_4 UL + U^2 = 0,$$

where $A = 0$, $U = 0$ are cones whose vertex is the node, and $L = 0$ is a plane through the node; we may write this equation

$$(U + x_4 L)^2 = x_4^2 (L^2 - A),$$

hence the node is the vertex of a cone of Kummer.

This result may also be seen from the fact that any tangent plane drawn to the surface from the node meets the surface in a quartic curve with four nodes, and if the surface is not ruled this section must consist of two conics.

The sextic tangent cone whose vertex is the node D, here consists of the cone V of Kummer of vertex D and the cone U (counted twice); the latter cone meets the surface in the double conic and in the four lines given by $A = U = 0$.

The surface contains twelve lines; for if the foregoing four lines meet the double conic in $P_1 \ldots P_4$, through the line DP we can draw two tangent planes to V each of which meets the quartic surface in two conics, and in each plane there is therefore one other line in addition to DP_1: similarly for the tangent planes drawn to V through the lines DP_2, DP_3, DP_4. Hence we have in all $4 + 8 = 12$ lines on the surface.

There are only three cones of Kummer in addition to V, for if we take the vertices of the triangle self-polar for the sections of

U and V by x_4 as vertices of reference, the equation of the surface may be written

$$\{ax_1^2 + bx_2^2 + cx_3^2 + 2x_4(\alpha x_1 + \beta x_2 + \gamma x_3) + 2\lambda x_4^2\}^2$$
$$= 4x_4^2\{x_1^2(1+\lambda a) + x_2^2(1+\lambda b) + x_3^2(1+\lambda c)$$
$$+ 2\lambda x_4(\alpha x_1 + \beta x_2 + \gamma x_3) + \lambda^2 x_4^2\};$$

and the values of λ for which the quadric on the right is a cone are given by the cubic equation

$$1 = \frac{\alpha^2}{1+\lambda a} + \frac{\beta^2}{1+\lambda b} + \frac{\gamma^2}{1+\lambda c}.$$

If there is a second node D', then if the cone V contains D' it will have a double edge and therefore consist of two planes, and the equation of the surface is

$$U^2 = 4w^2pq.$$

If V does not contain D' then U must contain it, and since the line DD' meets the double conic it therefore lies on the surface. The equation of the surface may be written in either of the forms

$$U^2 = 4w^2V, \quad U'^2 = 4w^2V',$$

where D, the vertex of V, lies upon $U=0$, and D', the vertex of V', lies upon $U'=0$.

In this case two of the lines $DP_1 \ldots DP_4$ must coincide, since otherwise D' could not be a double point of the curve of intersection of U and the quartic surface, consisting of four lines. In fact the tangent plane at any point of DD' meets the surface in a section which contains four nodes lying on DD', hence the section consists of the line DD' taken doubly together with a conic. The tangent plane at *any point* of DD' is the same since otherwise this line would be a double line of the surface. The line is torsal.

If there are three nodes the section of the surface through these nodes contains five double points and therefore consists of two lines and a conic; *one* line joining a pair of nodes does not lie on the surface, whose equation may be written in either of the forms

$$(w^2 + V - pq)^2 = 4w^2V, \quad (V - pq - w^2)^2 = 4w^2pq,$$

where the vertex of V lies upon $w^2 - pq = 0$. The lines joining the vertex of V to the two nodes each meet the double curve, hence the plane through these two lines meets the surface in each of them *doubly*.

If there is a fourth node *two* lines joining a pair of nodes do not lie on the surface, and four lines joining pairs of nodes lie on the surface, whose equation is therefore

$$(w^2 + rs - pq)^2 = 4w^2rs.$$

If the nodes on the lines (p, q) and (r, s) are D_1, D_2, and D_3, D_4 it is clear that the lines D_1D_3, D_1D_4, D_2D_3, D_2D_4 lie on the surface.

There cannot be more than four nodes, for if D be the node which is the vertex of a cone V, then V cannot contain more than one other node, hence the remaining nodes must lie on U, and it was seen that each node on U causes the coincidence of a pair of the lines $DP_1 \dots DP_4$: hence U cannot contain more than two nodes of the quartic surface apart from its vertex D*.

39. Special positions of the base-points.

It will now be shown that singularities of the surface arise from special relative positions of the base-points. If a node exists, any line through it meets the surface in two points apart from the node, hence any two cubic curves f_i which correspond to plane sections of the surface through the node meet in two variable points only.

In order that this may be possible *one* of the two following cases must arise: Either, in the first case, these cubics must have a common node and intersect in three other fixed points, e.g. if

$$\rho x_1 = \xi_3^2 A + \xi_3 L_1 + \Sigma_1, \quad \rho x_2 = \xi_3 L_2 + \Sigma_2,$$

$$\rho x_3 = \xi_3 L_3 + \Sigma_3, \quad \rho x_4 = \xi_3 L_4 + \Sigma_4,$$

where the L_i are quadratic in ξ_1, ξ_2 and the Σ_i cubic in ξ_1, ξ_2. The point (1000) will then be a node to which the point $\xi_1 = \xi_2 = 0$ will correspond. The system of cubic curves will touch at the point $\xi_1 = \xi_2 = 0$, so that two base-points coincide. Thus *the coincidence of two base-points leads to a node on the quartic surface.*

The four lines through the node correspond to the following: the point consecutive to $\xi_1 = \xi_2 = 0$ upon $A = 0$, and the joins of this point to the three other base-points.

Or, in the second case, three base-points are collinear, and we may take as equations of Clebsch

$$\rho x_1 = f_1, \quad \rho x_2 = \alpha u_2, \quad \rho x_3 = \alpha u_3, \quad \rho x_4 = \alpha u_4,$$

where $u_2 = 0$, $u_3 = 0$, $u_4 = 0$ are conics having two common points, which also lie upon $f_1 = 0$. The base-points are then given by $f_1 = \alpha = 0$ and the two other common points of the system.

* These surfaces have been investigated by Korndörfer, *Die Abbildung einer Fläche vierter Ord.*, etc., Math. Ann. I. and II.

The point (1000) is seen to be a node, for if

$$\frac{x_2}{A}=\frac{x_3}{B}=\frac{x_4}{C}$$

be any line through it, the points in which this line meets the surface have as their images the intersections of the conics

$$\frac{u_2}{A}=\frac{u_3}{B}=\frac{u_4}{C},$$

which are two in number, apart from the two base-points.

Hence *if three base-points are collinear the surface has a node.* The image of the node is here the line $\alpha=0$.

Two nodes on the surface may arise in three ways: first if the base-points are doubly collinear, e.g. when the join of the base-points 1, 5 meets the join of the base-points 2, 4 in the point 3; secondly when two base-points are coincident and three are collinear; thirdly if there is a double coincidence of two base-points.

Considering the first case, let $\alpha=0$, $\beta=0$ be the lines (1, 5) and (2, 4); the equations of Clebsch are here

$$\rho x_1=\alpha u,\quad \rho x_2=\alpha\beta L_1,\quad \rho x_3=\alpha\beta L_2,\quad \rho x_4=\beta v,$$

where $u=0$, $v=0$, are conics through two base-points; $L_1=0$ and $L_2=0$ are any lines.

Thus as in the case of one node the points A_1, A_4 are nodes, the line β corresponds to A_1 and α to A_4; to the point $\alpha=\beta=0$ corresponds the line A_1A_4 *which lies on the surface.*

There are nine lines on the surface whose images are the base-points and the lines 12, 45, 14, 25; those which correspond to the base-points 1, 3, 5 pass through one node and those to 2, 3, 4 through the other node.

There are three sets of pairs of conics: first those which have as their images the pencils of lines whose centres are the base-points 1 and 5, these conics pass through a node of the surface which is the vertex of a cone of Kummer; secondly those corresponding to the pencils of lines whose centres are 2 and 4, these conics also pass through a node which is the vertex of a cone of Kummer: lastly those which are represented by the conics through the base-points 1, 2, 4, 5 and the pencil of lines whose centre is 3.

That there are only three cones of Kummer may also be seen thus: referring to the equation of Art. 38; if D is the vertex of the cone

$$x^2+y^2+z^2=0,$$

and D' the vertex of

$$x^2(1+\lambda a)+y^2(1+\lambda b)+z^2(1+\lambda c)+2w\lambda(\alpha x+\beta y+\gamma z)+\lambda^2w^2=0,$$

the line DD' is

$$\frac{x(1+\lambda a)}{\alpha}=\frac{y(1+\lambda b)}{\beta}=\frac{z(1+\lambda c)}{\gamma},$$

and since DD' meets the double conic we have

$$\frac{a\alpha^2}{(1+\lambda a)^2}+\frac{b\beta^2}{(1+\lambda b)^2}+\frac{c\gamma^2}{(1+\lambda c)^2}=0,$$

which is the condition that the cubic equation for λ should have a pair of equal roots.

40. When three base-points are collinear and two are coincident the appropriate equations are

$$\rho x_i = a_i\xi_1^2\xi_3 + \xi_1 L_i + b_i S, \qquad (i=1, 2, 3, 4),$$

where the L_i are quadratic and S cubic in ξ_2, ξ_3.

Here the node b_i corresponds to the line $\xi_1=0$, and the node a_i to the point $\xi_2=\xi_3=0$. Thus the join of the nodes does not lie on the surface. The number of lines of the surface is easily seen to be eight. There are three cones of Kummer, as is seen by forming the discriminant of

$$pq+\lambda U+\lambda^2w^2,$$

the surface being $U^2=4w^2pq$.

When a coincidence of two base-points occurs twice we obtain the same surface as in the first case, for, as before, if a and b are the points at which coincidence occurs, each of these points will correspond to a node of the surface, and the line joining them is the image of a line on the surface.

The case of three nodes arises when the join of two consecutive base-points, say 1 and 5, passes through another base-point, say 3, the points 2 and 4 being coincident. The three nodes correspond to the line (1, 5) and to the points 1 and 2. There are six lines on the surface.

When the join of the (coincident) points 2 and 4 passes through 3 we have four nodes. The only lines on the surface are the four which respectively correspond to the point 3, the line 12 and the base-points consecutive to 1 and 2.

41. Coincidence of more than two base-points.

When three of the cubics have a common node and a common tangent thereat, e.g. if

$$f_i = \xi_3 \alpha L_i + \beta\gamma P_i \qquad (i = 1, 2, 3),$$

where the L_i, P_i, α, β, γ represent lines through the point

$$\xi_1 = \xi_2 = 0;$$

this common node counts as five intersections of any two of these three cubics. If S be any cubic passing through the points (ξ_3, β), (ξ_3, γ) and having α as its tangent at (ξ_1, ξ_2), the system of cubics

$$S + \sum_1^3 \lambda_i f_i = 0$$

intersect in three consecutive points at (ξ_1, ξ_2) and pass through two other fixed points. Here, therefore, *three* base-points are coincident.

Among the curves of the system appear (i) $\beta\gamma P = 0$, where P is a line through the point $\xi_1 = \xi_2 = 0$; and (ii) $\alpha(\xi_3 \alpha + c\beta\gamma) = 0$. Each of these curves is intersected by any curve f_i in one point only apart from the base-points; hence the node is *biplanar* with the planes (say x_2, x_3) corresponding to (i) and (ii) as tangent planes thereat.

When the nodal cubics f_i have three consecutive points common at (ξ_1, ξ_2) on the branch whose tangent is α, and one other common point, they are of the form

$$\xi_3 \alpha (p_i \beta + \alpha) + \gamma (q_i \beta^2 + r_i \alpha\beta + s_i \alpha^2) = 0;$$

where we have p_i/q_i the same for each cubic.

Hence the preceding cubic (i) is $\alpha\gamma P = 0$ and the planes x_2, x_3 intersect in the line of the surface given by $\alpha = 0$: the binode is therefore of the *second species*.

If $S = 0$ is a cubic through the three consecutive points and the additional point, the cubics $S + \sum_1^3 \lambda_i f_i = 0$ intersect in four consecutive points at (ξ_1, ξ_2); here, therefore, four base-points are coincident.

Lastly when the nodal cubics f_i have four consecutive points common at (ξ_1, ξ_2) on the branch whose tangent is α, they are of the form

$$\xi_3 \alpha (p_i \beta + \alpha) + q_i \beta^3 + r_i \beta^2 \alpha + s_i \beta\alpha^2 + t_i \alpha^3 = 0,$$

with the conditions, $q_i = A p_i$, $r_i = B + C p_i$.

Hence the curve (i) is $\alpha^2 P$, and the plane x_2 therefore touches the quartic surface along the line (x_2, x_3). The binode is of the third species. The cubics $S+\sum_1^3 \lambda_i f_i=0$ have five consecutive points in common, and the base-points all coincide.

The equation of a surface with a binode is of the form

$$V^2-2Vx_1x_4+x_4^2x_2x_3=0,$$

where $V=0$ is a quadric cone whose vertex is the binode.

From consideration of the base-points the surface is seen to contain eight lines.

If there is a further node Q, then as in Art. 38, the line joining Q to the biplanar node P lies in the surface, and hence in one of the planes x_2, x_3, say x_2; also as before the lines of intersection of x_2 and V coincide, and hence x_2 touches both V and the surface along the line PQ.

Conversely if x_2 touches the surface along a line the surface has a further node on that line *.

42. Uniplanar node.

When four base-points coincide in one point A, and the fifth base-point lies on the tangent at A to the cubics of the system, we have a uniplanar node. For let S be any particular cubic of the system and β the tangent to S at A, then any cubic of the system is represented by

$$S+a_i\xi_3\beta^2+\beta P_i=0,$$

where $P_i=0$ is a pair of lines through A.

The line $\beta=0$ corresponds to the node; the equations

$$\rho x_i=\beta(\xi_3\beta+P_i), \qquad (i=1, 2, 3)$$

represent plane sections through the node.

One derivable equation is $\rho x=\beta^2\alpha$, and the line $x=0$, $x_i=0$, meets the surface in one point only; this holds *only* for the plane $x=0$, hence the node is *uniplanar*, with $x=0$ as its tangent plane.

43. Ruled surfaces.

If *all* the cubic curves of the system have a common node and one other common point, three of the base-points become indeterminate, viz. three of them come into coincidence with the fourth

* If x_3 also touches V we have a second node.

in an indeterminate manner. The equations of Clebsch are of the form

$$\rho x_i = \xi_3 L_i + \xi_1 M_i,$$

where L_i, M_i are quadratic in ξ_1, ξ_2 and where we must assume *two linear relations* between the L_i to secure that the four cubics have all the four consecutive points in common. We may therefore take as equivalents of these equations the following, viz.

$$\rho x_1 = \xi_3 L_1, \quad \rho x_2 = \xi_3 L_2 + \xi_1 M_2, \quad \rho x_3 = \xi_1 M_3, \quad \rho x_4 = \xi_1 M_4.$$

It follows that to each line of the pencil $\xi_2 = \lambda \xi_1$ there corresponds a line on the surface, which is therefore ruled. Since each cubic is nodal and has therefore zero deficiency the surface must possess a double line in addition to the double conic.

To the line $\xi_1 = 0$, however, there corresponds the line $x_3 = x_4 = 0$, which is such that any plane through it meets the surface in two lines; hence the line (x_3, x_4) is the double line; through each point of the double line there pass two generators, viz. those obtained by giving any constant value to L_1/L_2.

If the fifth base-point coincides in a *definite way* with the point in which the other four base-points become coincident, the equations of Clebsch are of the form

$$\rho x_i = \xi_3 \alpha L_i + P_i Q_i R_i.$$

Hence if we join any point on the line p determined by the equation $\rho x_i = L_i$, to the corresponding point on the cubic curve given by the equations $\rho x_i = P_i Q_i R_i$, we obtain a generator of the surface; hence through each point of p, the double line, there passes *one* generator of the surface.

44. Cuspidal double curve.

We now consider special cases of the quartic surface with a nodal conic arising from peculiarities of the double curve. Taking the surface to be $x_1^2 V + U^2 = 0$, we obtain the two tangent planes at any point of the double curve by writing $x_i + \xi_i$ for x_i in this equation and selecting the terms of the second order in the ξ_i: this gives as their equation $\xi_1^2 V + (\Delta U)^2 = 0$, where

$$\Delta U \equiv \Sigma \xi_i \frac{\partial U}{\partial x_i}.$$

These planes coincide at each point of the double curve when U is a cone of which $x_1 = 0$ is a tangent plane; the double curve is then a line which is said to be *bidouble* (Segre); the corresponding equation of the quartic surface being then

$$x_1^2 U + (x_1 x_3 + x_2^2)^2 = 0.$$

This may be written in the form

$$x_1^2 V + x_1 x_2^2 A + x_2^4 = 0.$$

There are two triple points on the bidouble line, viz. those given by

$$x_1 = x_2 = V = 0.$$

The sections through this line consist of conics passing through the triple points which in a number of cases reduce to a pair of lines. Take the plane x_2 as containing such a pair and the tangent planes to V at the triple points as the planes x_3, x_4; the equation of the surface is then

$$x_1^2 (x_3 x_4 + a x_2^2 + b x_1 x_2) + x_1 x_2^2 A + x_2^4 = 0;$$

which may also be written

$$(x_3 x_1 - a_4 x_2^2)(x_4 x_1 - \beta_4 x_2^2) = x_2 \prod_1^3 (a_i x_1 + \beta_i x_2).$$

This shows that there are four planes through the bidouble line which contain a pair of lines of the surface.

Again the tangent planes will coincide at each point of the double curve for the case in which V contains the double curve: the surface is then

$$x_1^2 (x_1 x_2 + a U) + U^2 = 0,$$

which may be reduced to the form

$$x_1^3 x_2 + U^2 = 0.$$

The tangent planes at each point of the cuspidal double conic also touch U and hence meet in the pole of U for the plane x_1. The plane x_2 is a trope.

The surface has two "close-points" C, C', viz. those given by $x_1 = x_2 = U = 0$. Taking the planes δ, δ' which touch U at C and C' as the planes x_3, x_4, U takes the form

$$x_3 x_4 + (x_1, x_2 \between a)^2.$$

It is clear that the planes δ, δ' each contain four lines of the surface, those in δ passing through C, those in δ' passing through C'.

45. Involutory properties: class of the surface*.

Let us take any point x in CC' and its polar plane σ for U, σ being then the plane $x_4 \xi_3 + x_3 \xi_4 = 0$; so that if X be any point

* See Béla Totössy, *Ueber die Flächen vierter Ordnung mit Cuspidalkegelschnitt*, Math. Ann. XIX. (1882).

of σ the line $\lambda x_i + X_i$ meets the surface in the points given by the equation

$$\lambda^2 x_3 x_4 + U_X = \pm\sqrt{-X_1^3 X_2}.$$

Denoting these points by P, P'; Q, Q' it is clear that we thus obtain two sets of four harmonic points, viz., x, X, P, P' and x, X, Q, Q'; hence *the surface is in involutory central collineation with itself for any point K of CC' as centre, and with the polar plane of K for U as plane of collineation.* From consideration of a quadric which touches the surface at P and P', it is clear that the tangent planes at P and P' meet in a line of σ. If the line through x touches the surface, the points P, P' and X all coincide; hence the point of contact of any tangent line to the surface through x lies on a plane section of the surface; any such section is of class six since it possesses two cusps. Now the class of a tangent cone is equal to the class of its plane section, which is in this case *six*; and the complete tangent cone from x to the surface consists of the plane x_2, the plane x_1 taken thrice, and a quartic cone of class six; hence *six* must be the class of the tangent cone from *any* point to the surface; the surface is therefore of class six.

46. Cuspidal conic and additional node.

Any plane through the line CC' meets the surface in a pair of conics touching at C and C'; if there exists a node D of the surface outside the cuspidal conic, one of these conics must reduce to the lines DC, DC' which touch the residual conic of the section by the plane (DCC') at C and C' Two of the lines in δ coincide and pass through D, similarly for δ', hence the surface is a special case of those represented by the equation

$$(x_1 - kx_2)^2 (x_1, x_2 \between a)^2 + x_3 x_4 \{x_3 x_4 + (x_1, x_2 \between b)^2\} = 0.$$

The planes δ, δ' touch the surface along the lines DC, DC' respectively and cut it also along two pairs of lines passing respectively through C and C' and meeting on the line (δ, δ').

The equation of the tangent cone of the surface at D is

$$(x_1 - kx_2)^2 (k, 1 \between a)^2 X_2^2 + x_3 x_4 V_X = 0,$$

where X_i is the point D; if the node is biplanar we must have $(k, 1\between a)^2 = 0$, and the equation of the surface reduces to

$$(x_1 - kx_2)^3 L + x_3 x_4 \{x_3 x_4 + (x_1, x_2 \between b)^2\} = 0,$$

where $L = 0$ is a plane through C, C'.

Hence the planes δ, δ' osculate the surface along the lines CD, CD' and each contains one other line of the surface.

47. Double conic consisting of two lines.

When the double conic is a pair of lines the surface is represented by an equation of the form

$$(x_3x_4 + x_1A)^2 = x_1^2V,$$

$V = 0$ being a cone; if x_2 is any tangent plane of V, the foregoing may be written

$$(x_3x_4 + x_1A)^2 = x_1^2(C^2 + x_2B).$$

Let x_2 be one of the tangent planes of V which meets the surface in a pair of lines and a conic; one of these lines will meet one double line and the other will meet the second double line. Taking these four lines as edges of the tetrahedron of reference and expressing that the lines (x_2, x_3), (x_2, x_4) lie on the surface, the equation of the latter may be written in the form

$$(x_3x_4 + x_1\alpha)^2 = x_1^2(\alpha^2 + x_2\beta).$$

If any point on a double line be joined to any point on a simple line of the surface, this join meets the surface in one further point; this affords a means of representation of the surface on a plane; for if we write

$$x_4 = \frac{\xi_3}{\xi_1}x_1, \quad x_2 = \frac{\xi_2}{\xi_1}x_3,$$

the equation of the surface shows that

$$\rho x_1 = \xi_1 v, \quad \rho x_2 = \xi_2 u, \quad \rho x_3 = \xi_1 u, \quad \rho x_4 = \xi_3 v;$$

where $u = 0$, $v = 0$ are conics such that one of their intersections is the point (ξ_2, ξ_3), and where u passes through the point (ξ_1, ξ_3). The five base-points consist of the points (u, v) and the point (ξ_1, ξ_3).

The case of additional nodes arises as in the case of the surface with a nodal conic.

Either or both of these lines may be cuspidal. The equation of the surface in the latter case is

$$\{x_3x_4 - x_1(ax_1 + bx_2)\}^2 = x_1^3x_2.$$

An additional node exists if $4ab = 1$.

48. Classification of quartic surfaces with a nodal conic.

The method of Segre (Art. 35) affords a means for the classification of quartic surfaces with a nodal conic. The two four-dimensional varieties $F = 0$, $\Phi = 0$ are reduced by the method of Elementary Factors* of Weierstrass to their canonical forms,

* See *Quadratic Forms*, etc., Bromwich, Camb. Math. Tracts; or the Author's *Treatise on the Line Complex*.

leading to various types and each type to sub-cases. Each pair of forms thus arising affords one species of the quartic surface considered.

An elementary factor $(\lambda - \lambda_i)^{e_p}$ of $F + \lambda\Phi$ gives rise, if e_p is greater than unity, to a group of terms in e_p variables, viz.,

$$\lambda_i (x_1 x_{e_p} + x_2 x_{e_p - 1} + \ldots + x_{e_p} x_1) + x_1 x_{e_p - 1} + \ldots + x_{e_p - 1} x_1,$$
$$-(x_1 x_{e_p} + \ldots + x_{e_p} x_1),$$

in F and Φ respectively, so that $F + \lambda_i \Phi$ is a cone of the pencil (F, Φ) whose vertex lies on each variety of the pencil; at this point the varieties have a common tangent hyperplane x_1; this point is therefore a double point of Γ.

We now consider the principal types, indicating them as in Segre's notation by

$$\{11111\}, \quad \{2111\}, \quad \{221\}, \quad \{311\}, \quad \{23\}, \quad \{41\}, \quad \{5\}.$$

The surface which is the projection of Γ is denoted by [11111], etc., but if the point of projection A lies on a cone of (F, Φ) the projected surface is represented by $[\bar{1}1111]$, and so on.

The general type [11111] has been already considered in the preceding chapter; we may find its class by aid of this method.

Taking $$F \equiv \sum_1^5 x_i^2, \quad \Phi \equiv \sum_1^5 a_i x_i^2,$$

the required class is equal to the number of tangent planes of the projected surface which can be drawn through any line p of S_3; but if the projection of a plane π from A on S_3 passes through p, then A, p and π must lie in the same hyperplane; our problem is therefore to find the number of hyperplanes through the plane (A, p) which contain tangent planes of Γ. If the plane (A, p) is given by the equations

$$\sum_1^5 A_i x_i = 0, \quad \sum_1^5 B_i x_i = 0,$$

the condition requires that

$$\Sigma\, \alpha_i \xi_i \equiv \rho \Sigma\, \xi_i \frac{\partial F}{\partial x_i} + \sigma \Sigma\, \xi_i \frac{\partial \Phi}{\partial x_i},$$

where $\alpha_i = A_i + \lambda B_i$. Thus we obtain

$$\alpha_i = x_i (\rho + \sigma a_i), \qquad (i = 1, 2, 3, 4, 5),$$

whence
$$\Sigma \frac{\alpha_i^2}{\left(1+\frac{\sigma}{\rho} a_i\right)^2} = 0, \quad \Sigma \frac{a_i \alpha_i^2}{\left(1+\frac{\sigma}{\rho} a_i\right)^2} = 0,$$

and therefore
$$\Sigma \frac{\alpha_i^2}{1+\frac{\sigma}{\rho} a_i} = 0.$$

These equations show that the last equation, considered as a quartic in $\frac{\sigma}{\rho}$, has equal roots; and forming its discriminant, which is of degree six in α_i^2, we obtain an equation of degree twelve in λ. The class of the projected surface is therefore twelve.

We now proceed to consider the remaining six principal types. The canonical forms corresponding to the type {1112} are

$$F \equiv x_1^2 + x_2^2 + x_3^2 + 2x_4x_5,$$

$$\Phi \equiv a_1x_1^2 + a_2x_2^2 + a_3x_3^2 + 2a_4x_4x_5 + x_4^2.$$

At the point (00001) which is a double point of Γ, F and Φ have the common tangent hyperplane $x_4 = 0$. The tangent cone of Γ at this point is

$$x_4 = (a_1 - a_4)\, x_1^2 + (a_2 - a_4)\, x_2^2 + (a_3 - a_4)\, x_3^2 = 0.$$

This cone contains the four lines

$$x_4 = x_1^2 + x_2^2 + x_3^2 = a_1x_1^2 + a_2x_2^2 + a_3x_3^2 = 0,$$

which also belong to Γ.

If there is any additional line on Γ, the plane through it and the vertex of the cone $\Phi - a_4F$ must lie on this cone. Now through any generating line of a cone in S_4 we can draw two generating planes (one of each set), Art. 35, and hence through each of the preceding four lines; such a plane meets Γ in a conic which therefore reduces to two lines. Hence corresponding to each of the four lines through the double point we have two other lines of Γ. Therefore the surface [1112] has a conical node with four lines passing through it, and eight other lines. The class of the surface is ten, being diminished by two from that of [11111] owing to the additional node.

In [$\bar{1}$112] the same applies, the double conic being here two intersecting lines.

In [111$\bar{2}$] the point of projection lies on the cone $\Phi - a_4F$; the double point of Γ is projected into the intersection of the two double lines, and this point is now triple*; the tangent cone at it consists of the plane of the double lines and the projection of the tangent cone of Γ at its node.

If Γ is of the type $\{122\}$, F and Φ have the forms

$$F \equiv x_1^2 + 2x_2x_3 + 2x_4x_5,$$

$$\Phi \equiv a_1x_1^2 + 2a_2x_2x_3 + 2a_3x_4x_5 + x_2^2 + x_4^2.$$

Here Γ has two nodes, viz. (00100), (00001); the tangent cones thereat being

$$x_2 = (a_1 - a_2)\, x_1^2 + 2\,(a_3 - a_2)\, x_4x_5 + x_4^2 = 0;$$

and

$$x_4 = (a_1 - a_3)\, x_1^2 + 2\,(a_2 - a_3)\, x_2x_3 + x_2^2 = 0.$$

The line joining the nodes belongs to Γ, and along this line the plane $x_2 = x_4 = 0$ touches both F and Φ and therefore Γ.

Through the first point there pass the two lines

$$x_2 = x_1^2 + 2x_4x_5 = a_1x_1^2 + 2a_3x_4x_5 + x_4^2 = 0;$$

similarly two lines pass through the second point.

As in the case $\{1112\}$ each of these additional four lines gives rise to a line of Γ; hence Γ contains nine lines in all.

The nature of the surfaces [122], $[\bar{1}22]$, $[1\bar{2}2]$ is therefore determined.

For the type $\{113\}$ we have

$$F \equiv x_1^2 + x_2^2 + x_4^2 + 2x_3x_5,$$

$$\Phi \equiv a_1x_1^2 + a_2x_2^2 + a_3\,(x_4^2 + 2x_3x_5) + 2x_3x_4.$$

The point (00001) is a double point of Γ; the tangent cone at it, which is represented by

$$x_3 = 0, \quad (a_1 - a_3)\, x_1^2 + (a_2 - a_3)\, x_2^2 = 0,$$

breaks up into two planes μ_1, μ_2, whose intersection does not lie on Γ. The point is therefore *biplanar*. It is easily seen that through the double point there pass four lines of Γ, of which two r_1, r_1' lie in μ_1 and two r_2, r_2' in μ_2. Through r_1 there passes a generating plane of the cone $\Phi - a_3F = 0$, of the same system as μ_2;

* For any plane α passing through A and the vertex K of the cone corresponding to 2 meets that cone in two lines, and each of these meets Γ in one other point giving two points Q, R of Γ on α. The plane α is projected from A on S_3 into a line r passing through K' the projection of K; and Q, R are projected into the two other points in which r meets the surface. If however A lies on the cone, one of the two previous lines must pass through A, and r thus meets the surface in one point only (apart from K'); the point K' is therefore *triple*.

and so for the three other lines r_1', r_2, r_2'; hence each of these four planes meets Γ in an additional line, giving rise to four new lines s_1, s_1', s_2, s_2'.

Hence applying to the surface [113] we have a surface of the ninth class* which has a biplanar node and contains eight lines.

For the type {23} we have

$$F = 2x_1x_2 + 2x_3x_5 + x_4^2,$$

$$\Phi = 2a_1x_1x_2 + x_1^2 + a_2(2x_3x_5 + x_4^2) + 2x_3x_4.$$

From consideration of the cases {1112}, {113} it is seen that Γ possesses a conical node at D and a biplanar node at D' at which the tangent cone breaks up into two planes μ_1 and μ_2. The line DD' is given by $x_1 = x_3 = x_4 = 0$; and the plane μ_1 is $x_1 = x_3 = 0$; this touches Γ along the line DD'. As in {113} there are two lines r_2, r_2' in the plane μ_2.

The section of Γ by its tangent hyperplane x_1 at D is

$$x_1 = 2x_3x_5 + x_4^2 = x_3x_4 = 0;$$

and is therefore the line DD' together with one other line. The two generating planes of the cone whose vertex is D which pass through the latter line, meet the surface in two new lines. These six lines constitute all the lines of the surface.

The nature of the surfaces [23], [2$\bar{3}$], [$\overline{23}$] follows immediately; they are of the seventh class, the first has a conical node and a biplanar node, the second has a conical node and a triplanar point, the third has a biplanar node and the intersection of the double lines as a triplanar point.

For the type {14} we have

$$F \equiv x_1^2 + 2x_2x_5 + 2x_3x_4,$$

$$\Phi \equiv a_1x_1^2 + 2a_2(x_2x_5 + x_3x_4) + 2x_2x_4 + x_3^2.$$

The double point D, or (01000), is here biplanar, and the two nodal planes intersect in a line which lies on Γ; the biplanar point is therefore of the second kind†.

The nodal planes meet Γ also in two lines r_1, r_2 through D. In the two other generating planes of the cone whose vertex is D which pass through r_1 and r_2 respectively there are two other lines

* A biplanar node of the first kind reduces the class of the surface by three. Salmon, *Geom. of three dimensions*, p. 489.

† See Salmon, *l.c.*

of Γ (say) s_1 and s_2, which meet r_1 and r_2 respectively; and since the line of intersection of these latter planes meets s_1 and s_2 it must therefore meet them in the same point. Hence we have four lines on Γ forming a skew quadrilateral, together with another line through D.

The preceding defines the surfaces [14], [$\overline{1}4$], [$1\overline{4}$] which are of the eighth class.

For the type $\{5\}$ we have

$$F \equiv 2x_1x_5 + 2x_2x_4 + x_3^2,$$
$$\Phi \equiv a_1(2x_1x_5 + 2x_2x_4 + x_3^2) + 2x_1x_4 + 2x_2x_3.$$

The (one) cone of the pencil meets x_1 in the two planes

$$x_1 = x_2 = 0; \quad x_1 = x_3 = 0;$$

and these planes meet in a line r of Γ. The first plane touches Γ along r; thus since one of the nodal planes touches Γ along r, the biplanar point is of the third species.

The other nodal plane meets Γ in a line r', through D. Another generating plane of the cone passes through r' which meets Γ in a line s. The lines r, r' and s are the only lines on Γ.

The properties of the surfaces [5] and [$\overline{5}$] follow; they are of the seventh class; the latter has two double lines meeting in a triplanar point.

49. Cones of the second species.

In the preceding types the pencil (F, Φ) contains cones, the equation of each cone being expressible in terms of four variables. When, however, two elementary factors are equal, the equation of the corresponding cone $\Phi - a_iF = 0$ contains not more than three variables and the cone is said to be of the second species; e.g. in $\{(11)\,111\}$ we have

$$\Phi - a_1F \equiv (a_3 - a_1)x_3^2 + (a_4 - a_1)x_4^2 + (a_5 - a_1)x_5^2.$$

This cone has ∞^1 generating planes through the line

$$x_3 = x_4 = x_5 = 0.$$

In the previous types there were seen to be two systems of generating planes given as the intersection of a cone with its tangent hyperplanes. In the present case we have one system of generating planes obtained as the intersection of the cone with its tangent hyperplanes which all pass through the line

$$x_3 = x_4 = x_5 = 0,$$

the *edge* of the cone. In each generating plane there is one conic of Γ.

Each of these conics passes through the two points of intersection of the *edge* of the cone with Γ. At either of these points there is a tangent hyperplane common to the pencil; these points are therefore double points of Γ.

If the group considered is (11) there are two such double points; if it is (21), (31), (41) the points coincide, as is seen by reference to the corresponding forms.

In the cases $\{1(22)\}$, $\{(23)\}$ the edge itself is seen to lie on Γ; in these cases since each generating plane of the cone meets Γ in the edge and one other line, there arise ∞^1 lines on Γ, which is therefore a ruled surface, having the edge as a double line.

We can easily determine the number of lines on Γ in the other cases; for any line of Γ must lie on this cone of the second species and therefore meet the edge in one of the double points of Γ upon it. The tangent hyperplane at either of these double points meets the pencil (F, Φ) in a pencil of ordinary quadric cones having the double point as vertex; the lines of intersection of two of these cones will be the lines of Γ through the point. Hence these surfaces cannot have more than eight lines.

Projecting Γ on S_3 gives us the surface we are investigating. In this case, however, the point of projection A may lie on a cone of the pencil (F, Φ) of the second species. Here only one generating plane of this cone passes through A, which cuts Γ in a conic which is projected from A on S_3 into a line which will be a double line of the projected surface. Reference to Art. 35 shows that if $f=0$ is a cone of the second species, the double line of the projected surface, given by $f = Df = \Sigma a_i x_i = 0$, is therefore to be regarded as arising from the coincidence of two double lines*. This line contains two *triple* points (distinct or coincident), the projections of the two double points of Γ which lie on the edge a of Γ. For the generating planes of f cut Γ in conics through the two double points, hence their projections from A meet S_3 in conics through the projections of these points which are therefore triple†.

Each of the ∞^2 hyperplanes through the edge of the cone meets the cone in two planes; each tangent hyperplane of the

* Segre calls this line *bidouble*, see page 70.

† Since any line through one of these points on the projected surface meets the surface in one other point only; see page 75, footnote.

cone meets it in a generating plane counted twice, and hence touches Γ in a conic. If $f=0$ is the variety of the pencil (F, Φ) which passes through A, and $c_x=0$ any hyperplane, the intersection of f and c_x is projected from A upon S_3 into the quadric

$c_x Df - c_{x'} f = 0,\ \sum_1^5 a_i x_i = 0$; where $\sum_1^5 a_i x_i = 0$ represents S_3. This quadric passes through the double conic $f = Df = 0$. Now let c_x be one of the preceding hyperplanes through the edge of the cone of the second order; we obtain on projection ∞^2 *quadrics through the double conic and the two double points; each meets the quartic surface in two conics.*

If c_x is one of the ∞^1 tangent hyperplanes of the cone of the second species we obtain on projection ∞^1 *quadrics touching the quartic surface along a conic and passing through the double conic.*

Through any point of S_4 there pass two of these tangent hyperplanes, hence through any point of S_3 there pass two quadrics containing the double curve which touch the quartic surface along a conic. Thus the quartic surface is the envelope of a system of quadrics simply infinite and of the second order which pass through the double conic and the two double points of the quartic surface*.

The existence of this set of quadrics is peculiar to those surfaces which have a cone of the second species. For such a quadric is the projection from A of the intersection of some hyperplane c_x with f, the variety through A. This hyperplane therefore touches Γ along a conic, and hence c_x meets the pencil (F, Φ) in a pencil of quadrics which touch along this conic; among these quadrics is therefore included the plane of the conic counted twice; if $c_x = d_x = 0$ is this plane, it is seen that among the varieties of the pencil (F, Φ) there is one of the form $d_x^2 + c_x e_x$, and this is a cone of the second species.

50. Quartic surfaces with a cuspidal conic.

It was seen (Art. 35) that if $f=0$ is the variety of the pencil (F, Φ) which passes through A (or x'), and ϕ any variety of the pencil, the equations of the projection of Γ from A on S_3 are

$$\{2f\phi' - Df D\phi\}^2 = (Df)^2 \{(D\phi)^2 - 4\phi\phi'\},$$
$$\Sigma a_i x_i = 0.$$

* For the surface $\phi^2 = 4w^2pq$, the quadrics are $\mu^2 wp + \mu\phi + wq = 0$. We have also, as in the general case, the quadrics $\lambda^2 w^2 + \lambda\phi + pq = 0$, which touch the surface along quadri-quartics.

Let x be any point on the double conic and $x+\xi$ a point on the surface contiguous to x; substituting $x+\xi$ for x and retaining only terms of the second order in ξ, we obtain as one of the equations of the two tangent planes to the surface at x,

$$\{2L\phi' - MD\phi\}^2 = M^2\{(D\phi)^2 - 4\phi\phi'\},$$

where L and M are the terms of the first order in ξ_i arising from f and Df respectively.

If these planes coincide we have

$$(D\phi)^2 - 4\phi\phi' = 0.$$

This equation together with $f=0$, $Df=0$, $\Sigma a_i x_i = 0$, gives the four pinch-points on the double conic. These planes coincide at each point of the double curve if the *tangent cone* to $\phi=0$ from x' contains the three-dimensional cone $f=0$, $Df=0$. Hence we have an identity of the form

$$4\phi\phi' - (D\phi)^2 \equiv Af + XDf,$$

i.e.

$$4\phi\phi' - Af \equiv (D\phi)^2 + XDf.$$

This shows that the pencil must contain a cone of the second species.

Thus having given a pencil (F, Φ) which contains a cone ψ of the second species, the surface Γ projected from A on S_3 has a cuspidal conic provided that A is so chosen that the tangent hyperplane at A of the variety through A is also a tangent hyperplane of ψ.

The equations of the surface given at the beginning of this article may therefore, when a cuspidal conic exists, be written in the form

$$\sum_1^5 a_i x_i = 0, \quad (2f\phi' - Df D\phi)^2 = (Df)^2\{Af + XDf\};$$

the latter equation is

$$(2f\chi' - Df D\chi)^2 = (Df)^3 L,$$

where $\chi = \phi + \dfrac{A}{4\phi'}f$, and L is linear in the variables. This is the equation obtained in Art. 44.

The close-points.

The two intersections of the edge of a cone of the second species with Γ were seen, in the general case, to give rise to two nodes on the projected surface; when a cuspidal conic exists, since the tangent hyperplane of the variety through P passes through the edge of this cone, these two intersections are therefore projected into two points on the cuspidal conic; they are the two *close-points.*

Quartic surfaces with a bidouble line.

If A (or x') lies on a cone of the second order $\psi=0$, then $D\psi=0$ touches ψ along a plane π, also π meets ϕ (any variety of the pencil) in a conic c^2 on Γ. The tangent plane to Γ at any point x of c^2 is given by the equations

$$\Sigma\xi_i\frac{\partial\psi}{\partial x_i}=0,\quad \Sigma\xi_i\frac{\partial\phi}{\partial x_i}=0.$$

If we suppose, as is permissible, ψ to be of the form $\sum_1^3 a_i\xi_i^2=0$, it is seen that the first of these hyperplanes is identical with $D\psi=0$, since for each point x of π we have

$$\frac{x_1}{x_1'}=\frac{x_2}{x_2'}=\frac{x_3}{x_3'}.$$

Hence the tangent plane of Γ at any point of c^2 lies in the fixed hyperplane $D\psi=0$, and is therefore projected from A into the same plane of S_3, viz. $D\psi=0$, $\Sigma a_i x_i=0$. The pair of tangent planes at each point of the bidouble line coincide.

51. Of the sub-types arising from the equality of elementary factors the first is

$$\{(11)\,111\}.$$

As stated in Art. 49 we have two nodes on Γ; the line joining them does not belong to Γ. Hence there arises the surface [(11) 111], treated in Art. 38, possessing two nodes whose join does not lie on the surface. This includes the special case of a cuspidal double conic.

Other special cases are $[\overline{1}\,(11)\,11]$, $[(\overline{11})\,111]$ having respectively two double lines and two nodes, and a cuspidal line containing two triple points.

The characteristics of the various other sub-types are given in the table at the end of this chapter*.

52. Steiner's surface.

The pencil (F, Φ) may consist entirely of cones of the first order having a common generator and a common tangent hyperplane along this generator.

* For many details see Segre, *loc. cit.*

Such a system, for instance, arises from the cones

$$\left.\begin{aligned} x_1x_4 + (a \between x_2x_3)^2 = 0 \\ x_1x_5 + (b \between x_2x_3)^2 = 0 \end{aligned}\right\} \quad \ldots\ldots\ldots\ldots\ldots\ldots(1).$$

The line upon which the vertices of these cones lie is

$$x_1 = x_2 = x_3 = 0,$$

this line is a double line of Γ. Through A, the point of projection, there passes one cone of the system, its two generating planes through A intersect on a line which meets the double line of Γ in the vertex of this cone. Hence the projected surface has *three* concurrent double lines, viz. the projection of the double line of Γ and the projections of the conics in the two generating planes through A.

Each of the ∞^1 cones has two sets of generating planes meeting Γ in conics, hence arise ∞^2 pairs of conics in plane sections of the projected surface. Three of the points of intersection of such a pair of conics lie on the three double lines, the fourth point is a point of contact of the plane with the surface. The surface is therefore a Steiner's surface (Chapter VII).

53. We add Segre's Table which contains a complete list of the different kinds of quartic surfaces with a double conic (including two lines or a bidouble line).

Index	Class of the surface	Character of the surface
[11111]	12	General surface
[2111]	10	One node
[311]	9	Biplanar point of the first species
[221]	8	Two nodes; the line joining them belongs to the surface
[41]	8	Biplanar point of second species
[32]	7	One node and a biplanar point of first species
[5]	7	Biplanar point of third species
[(11) 111]	8	Two nodes; the line joining them does not belong to the surface
[(21) 11]	8	A biplanar point of the second species
[(11) 21]	6	Three nodes; the lines joining two of them to the third belong to the surface
[(21) 2]	6	A node and a biplanar point of the second species
[(31) 1]	6	A uniplanar point of the first species
[(11) 3]	5	Two nodes and a biplanar point of the first species
[(41)]	5	A uniplanar point of the second species

Index	Class of the surface	Character of the surface
[(11) (11) 1]	4	Two pairs of nodes
[(21) (11)]	4	A pair of nodes and a biplanar point of the second species
[(22) 1]	4	Ruled surface (class II of Cremona)
[(32)]	4	Ruled surface (class IV of Cremona)

Surfaces with a cuspidal conic.

[(11) 111]	6	General case
[(21) 11]	6	The close-points of the double conic coincide
[(11) 21]	4	One node
[(21) 2]	4	The close-points coincide, one node
[(31) 1]	4	There is a point in which the two close-points coincide with a node
[(11) 3]	3	A biplanar point of the first species
[(41)]	3	A singular point of coincidence of the close-points with a biplanar point

Surfaces with two double lines (meeting in a point which is not a triple point).

[$\bar{1}$1111]	12	General case
[$\bar{1}$211]	10	One node
[$\bar{1}$31]	9	A biplanar point of the first species
[$\bar{1}$22]	8	Two nodes; the line joining them belongs to the surface
[$\bar{1}$4]	8	A biplanar point of the second species
[$\bar{1}$ (11) 11]	8	Two nodes
[$\bar{1}$ (21) 1]	8	A biplanar point of the second species
[1 (11) 2]	6	Three nodes
[$\bar{1}$ (31)]	6	A uniplanar point of the first species
[$\bar{1}$ (11) (11)]	4	Two pairs of nodes
[$\bar{1}$ (22)]	4	Ruled surface with three double lines

Surfaces with a double line and a cuspidal line.

[$\bar{1}$22]	8	General case
[$\bar{1}$4]	8	The close-points coincide
[$\bar{1}$ (11) 2]	6	One node
[$\bar{1}$ (31)]	6	The preceding node lies on the cuspidal line
[$\bar{1}$ (11) (11)]	4	Two nodes
[$\bar{1}$ (22)]	4	Ruled surface with two double lines and a cuspidal generator
[$\bar{1}$ (22)]	4	Ruled surface with two coincident directrices and a double generator

Surfaces with two cuspidal lines.

Index	Class of the surface	Character of the surface
$[\bar{1}\,(11)\,2]$	6	General case
$[\bar{1}\,(31)]$	6	Particular case
$[\bar{1}\,(11)\,(11)]$	4	One node

Surfaces with a triple point through which two double lines pass.

$[\bar{2}111]$	10	General case; the tangent cone at the triple point consists of the plane of the double lines and a quadric cone
$[\bar{3}11]$	9	The triple point is triplanar
$[\bar{2}21]$	8	One node
$[\bar{4}1]$	8	The triple point is a special triplanar point
$[\bar{2}3]$	7	A biplanar point of the first species
$[\bar{3}2]$	7	One node; the triple point is triplanar
$[\bar{5}]$	7	The triple point is a special triplanar point
$[\bar{2}\,(11)\,1]$	6	Two nodes
$[\bar{2}\,(21)]$	6	A biplanar point of the second species
$[\bar{3}\,(11)]$	5	The triple point is triplanar; there are two nodes

Surfaces with a triple point through which there pass a double line and a cuspidal line, or two cuspidal lines.

$[\bar{3}2]$	7	One double and one cuspidal line
$[\bar{5}]$	7	The close-point coincides with the triple point
$[\bar{3}\,(11)]$	5	One node
$[\bar{3}\,(11)]$	5	Two cuspidal lines

Steiner's surface.

	3	General case
	3	Two of the double lines coincide
	3	The three double lines coincide

Surfaces with a bidouble line (containing two triple points distinct or coincident).

$[(\bar{1}\bar{1})\,111]$	8	General case; the tangent cone at each triple point breaks up into a plane and a quadric cone
$[(\bar{2}\bar{1})\,11]$	8	The triple points coincide in a triplanar point
$[(\bar{1}\bar{1})\,21]$	6	One node
$[(\bar{2}\bar{1})\,2]$	6	The triple points coincide; one node
$[(\bar{1}\bar{1})\,3]$	5	A biplanar point of the first species
$[(\bar{4}\bar{1})]$	5	The double nodal plane of the triple point of the last case but one touches along a simple line
$[(\bar{1}\bar{1})\,(11)\,1]$	4	Two nodes
$[(\bar{1}1)\,(21)]$	4	A biplanar point of the first species
$[(\bar{2}\bar{1})\,(11)]$	4	The two triple points coincide; two nodes

Surfaces with a cuspidal line of the second species.

Index	Class of the surface	Character of the surface
$[(\overline{21})\ 2]$	6	General case; the cuspidal line contains a triple point and a point of osculation of the two sheets
$[(\overline{41})]$	5	The points just mentioned coincide in a triple triplanar point
$[(\overline{21})\ (11)]$	4	One node

Ruled surfaces with a triple line.

Index	Class of the surface	Character of the surface
$[(\overline{22})\ 1]$	4	General case (class III of Cremona)
$[(\overline{32})]$	4	Ruled surface (special case of class X of Cremona)

CHAPTER V

THE CYCLIDE

54. When the double conic is the section of a sphere by the plane at infinity we obtain the surface known as the *cyclide**. The equation of a cyclide in Cartesian coordinates is therefore $S^2+u=0$; where $S=0$ represents a sphere, and $u=0$ is a quadric.

Taking the centre of S as the origin and the axes in the directions of the principal axes of u, we obtain as the equation of the surface

$$(x^2+y^2+z^2)^2+4(A_1x^2+A_2y^2+A_3z^2+2B_1x+2B_2y+2B_3z+C)=0.$$

As in Chapter III we may write this equation in the form

$$(x^2+y^2+z^2-2\lambda)^2+4\{(A_1+\lambda)x^2+(A_2+\lambda)y^2+(A_3+\lambda)z^2 \\ +2B_1x+2B_2y+2B_3z+C-\lambda^2\}=0.$$

The second member of the left side will be a cone, $V=0$, if its discriminant is zero: this condition may be written in either of the forms

$$\phi(\lambda)\equiv C-\lambda^2-\left(\frac{B_1^2}{A_1+\lambda}+\frac{B_2^2}{A_2+\lambda}+\frac{B_3^2}{A_3+\lambda}\right)=0,$$

$$F(\lambda)\equiv(A_1+\lambda)(A_2+\lambda)(A_3+\lambda)(C-\lambda^2)-\{B_1^2(A_2+\lambda)(A_3+\lambda) \\ +B_2^2(A_3+\lambda)(A_1+\lambda)+B_3^2(A_1+\lambda)(A_2+\lambda)\}=0.$$

We thus obtain five values for λ, giving five cones. If one such cone V be $XY-L^2=0$, where $L=0$ is any plane through its vertex, the equation of the surface is

$$(x^2+y^2+z^2-2\lambda)^2+4(XY-L^2)=0 \qquad \text{.........(1)}.$$

As before (Chapter III) any tangent plane of the cone meets the surface in two circles, and every circle on the surface lies in a tangent plane to one of the five cones.

* For an extensive discussion of this surface see the work by Darboux entitled *Sur une classe remarquable de courbes et de surfaces quelconques.*

The intrinsic interest of this surface justifies a special discussion by use of Cartesian coordinates, showing the various real forms of the surface.

Again the sphere

$$x^2+y^2+z^2=2(L+\lambda) \quad \dots\dots\dots\dots\dots\dots(2)$$

meets the surface in a pair of circles lying on $X=0$, $Y=0$ respectively; the points of intersection of these circles being points of contact of the sphere and surface. Hence *the surface is the envelope of these bitangent spheres.* Moreover every bitangent sphere must arise in this manner; for if $x^2+y^2+z^2=2(\lambda+M)$ be a bitangent sphere it will meet the surface in a pair of circles and we may take the plane of one of them to be $X=0$, whence $M=\pm(L+kX)$, i.e. the surface may be written in the form

$$(x^2+y^2+z^2-2\lambda)^2+4(XY'-M^2)=0.$$

If $L=\alpha x+\beta y+\gamma z+\delta$, the condition that $L=0$ passes through the vertex of V gives

$$\frac{\alpha B_1}{A_1+\lambda}+\frac{\beta B_2}{A_2+\lambda}+\frac{\gamma B_3}{A_3+\lambda}-\delta=0;$$

and this is the condition that this bitangent sphere should cut orthogonally the sphere whose equation is

$$x^2+y^2+z^2+\frac{2B_1x}{A_1+\lambda}+\frac{2B_2y}{A_2+\lambda}+\frac{2B_3z}{A_3+\lambda}+2\lambda=0 \dots(3).$$

Again since

$$(A_1+\lambda)x^2+(A_2+\lambda)y^2+(A_3+\lambda)z^2 \\ +2B_1x+2B_2y+2B_3z+C-\lambda^2\equiv XY-L^2,$$

considering only terms of the second degree it follows that

$$(A_1+\lambda)x^2+(A_2+\lambda)y^2+(A_3+\lambda)z^2+(\alpha x+\beta y+\gamma z)^2$$

must break up into linear factors, hence

$$\frac{\alpha^2}{A_1+\lambda}+\frac{\beta^2}{A_2+\lambda}+\frac{\gamma^2}{A_3+\lambda}+1=0^* \quad \dots\dots\dots(4).$$

Hence *the cyclide may be generated in five ways as the envelope of a sphere whose centre lies on one of five fixed quadrics $Q_1 \dots Q_5$ and which cuts a fixed sphere† orthogonally.*

The quadrics Q_i are seen to be confocal. At each point of intersection of a quadric Q_i with the corresponding sphere S_i we have a bitangent sphere of zero radius; its centre is therefore a *focus* of the surface; hence arise five focal curves.

* The cone V is the reciprocal of the asymptotic cone of this quadric.

† This sphere is one of the quadrics H of Art. 31; its centre is the vertex of the cone V.

55. The five spheres $S_1 \dots S_5$ are mutually orthogonal; for the condition that any two of them, corresponding say to λ_1 and λ_2, should be orthogonal is

$$\frac{2B_1^2}{(A_1+\lambda_1)(A_1+\lambda_2)}+\frac{2B_2^2}{(A_2+\lambda_1)(A_2+\lambda_2)}$$
$$+\frac{2B_3^2}{(A_3+\lambda_1)(A_3+\lambda_2)}-2\lambda_1-2\lambda_2=0;$$

which follows at once from the equation

$$\phi(\lambda_1)-\phi(\lambda_2)=0.$$

Consideration of the equation $F(\lambda)=0$ shows that it has in general at least three real roots; since, taking $-A_1, -A_2, -A_3$ as in ascending order of magnitude, there lie an odd number of roots in each of the three intervals

$$-\infty \dots -A_1, \quad -A_1 \dots -A_2, \quad -A_2 \dots -A_3.$$

Hence there are in general at least three real pairs S_i, Q_i and there may be five.

Important relationships between the spheres S_i and the quadrics Q_i are the following: *the centres of any four of the spheres form a self-polar tetrahedron for the remaining sphere and for its corresponding quadric.* For expressing that the spheres corresponding to λ_1 and λ_3 are orthogonal we obtain an equation similar to the last; subtraction, and division by $\lambda_2-\lambda_3$ gives us

$$\frac{B_1^2}{(A_1+\lambda_1)(A_1+\lambda_2)(A_1+\lambda_3)}+\frac{B_2^2}{(A_2+\lambda_1)(A_2+\lambda_2)(A_2+\lambda_3)}$$
$$+\frac{B_3^2}{(A_3+\lambda_1)(A_3+\lambda_2)(A_3+\lambda_3)}+1=0,$$

which is the condition that the centre of the sphere S_3 should lie in the plane

$$\frac{B_1x}{(A_1+\lambda_1)(A_1+\lambda_2)}+\frac{B_2y}{(A_2+\lambda_1)(A_2+\lambda_2)}+\frac{B_3z}{(A_3+\lambda_1)(A_3+\lambda_2)}=1.$$

But the last equation represents the polar plane of the centre of S_2 with regard to Q_1. Similarly this plane passes through the centres of S_4 and S_5; and the centres of $S_2 \dots S_5$ form a self-polar tetrahedron with regard to Q_1.

Again representing any one of the spheres S_i by the equation

$$x^2+y^2+z^2+2f_ix+2g_iy+2h_iz+c_i=0,$$

we derive, from the fact that the spheres are mutually orthogonal, the equations

$$f_2 f_5 + g_2 g_5 + h_2 h_5 = \frac{c_2 + c_5}{2},$$

$$f_1 f_2 + g_1 g_2 + h_1 h_2 = \frac{c_1 + c_2}{2},$$

hence $$f_2(f_1 - f_5) + g_2(g_1 - g_5) + h_2(h_1 - h_5) = \frac{c_1 - c_5}{2},$$

which is the condition that the polar plane of the centre of S_5 for S_1, i.e.

$$-f_5(x + f_1) - g_5(y + g_1) - h_5(z + h_1) + f_1 x + g_1 y + h_1 z + c_1 = 0$$

or $$(f_1 - f_5)x + (g_1 - g_5)y + (h_1 - h_5)z + \frac{c_1 - c_5}{2} = 0,$$

should pass through the centre of S_2.

Similarly this plane passes through the centres of S_3 and S_4. Thus the tetrahedron formed by the centres of $S_2 \ldots S_5$ is self-polar for S_1.

56. Inverse points on the surface.

It is obvious from the form of its equation that the cyclide is inverted from any general point into another cyclide. If the centre of inversion be the centre of one of the *principal* spheres S_i, then since the surface is the envelope of spheres which cut S_i orthogonally, it is clear that the bitangent spheres are inverted into themselves (if the constant of inversion be the radius of S_i). Hence it follows that the two points of contact of a bitangent sphere of this system are collinear with the centre of S_i, and the surface is inverted into itself. This can also be seen as follows: the centres of the bitangent spheres in the neighbourhood of a point P of the quadric Q_i lie in the plane π tangent to Q_i at P, and these spheres all pass through the same two points M, M' of the cyclide; since S_i cuts all these spheres orthogonally its centre O must be collinear with M and M', and the line OMM' is perpendicular to the plane of their centres, i.e. π, and

$$OM \cdot OM' = R_i^2,$$

if R_i is the radius of S_i.

Thus M and M' are inverse points on the surface.

Again, all the spheres whose centres lie in π and which cut the sphere S_i orthogonally, will also cut orthogonally every sphere through the intersection of S_i and π, and in particular the two point-spheres which pass through the intersection of S_i and π. The centres of these point-spheres are therefore the points M and M'. *Hence the surface may be defined as the locus of the limiting points determined by S_i and the tangent planes to Q_i.*

The points of Q_i which give rise to real points of the cyclide are therefore those the tangent planes at which do not meet S_i in real points. Taking the tangent planes common to S_i and Q_i we have a curve or curves determined on Q_i defining the region on Q_i which gives rise to real points of the cyclide.

Bitangent spheres whose centres lie on the same generator of a principal quadric.

The spheres which cut S_i orthogonally and whose centres lie on a line p, a generator of Q_i, will pass through the points of contact P, P' of the tangent planes to S_i through p; hence if C is the point of intersection of p and a plane through the centre O of S_i perpendicular to p, each of these spheres will pass through the circle whose centre is C and radius CP (or CP'). The circle lies on the cyclide; for considering all the planes through p, the limiting points M, M' which arise in connection with S_i, lie in the plane of this circle, also $CM = CM' = CP = CP'$.

Hence real circles arise from those generators of Q_i which do not meet S_i in real points.

Taking all the planes through O perpendicular to each generator of the system to which p belongs we obtain ∞^1 sections of the cyclide consisting of two circles.

Conversely all the spheres which meet the cyclide in the same real circle will meet it again in circles and will be bitangent spheres; since their centres lie on the same real line, the quadric to which they belong must be a hyperboloid of one sheet; hence this type of quadric alone will give rise to real circles on bitangent spheres. We observe that of the three real quadrics Q_i which in all cases exist, *one* is an ellipsoid, *one* a hyperboloid of one sheet and *one* a hyperboloid of two sheets, corresponding respectively to the three real values of λ mentioned in Art. 54.

57. Roots of fundamental quintic. Focal curves.

It has already been seen (Art. 55) that in the general case in which $F(\lambda) = 0$ does not possess equal roots, it has an odd number of real roots in each of the intervals

$$-\infty \ldots -A_1, \quad -A_1 \ldots -A_2, \quad -A_2 \ldots -A_3,$$

where we suppose $-A_1$, $-A_2$, $-A_3$ arranged in ascending algebraic order of magnitude.

If three roots only are real then two are conjugate imaginary; it may be shown that any two corresponding real surfaces S_i, Q_i meet each other in a curve consisting of one portion only; for since three centres of spheres S_i are real and two conjugate imaginary, we may in three ways select a real pair S_i, Q_i so that their self-polar tetrahedron has two vertices real and two conjugate imaginary. If S_i, Q_i form such a pair, it may easily be seen that two of the four cones passing through their curve of intersection have equations of the form

$$x_1^2 + p\,(x_3^2 - x_4^2) + 2qx_3x_4 = 0,$$
$$x_2^2 + p'\,(x_3^2 - x_4^2) + 2q'x_3x_4 = 0.$$

Each generator of the first cone meets this curve in two points, which coincide if

$$p'\,(x_3^2 - x_4^2) + 2q'x_3x_4 = 0.$$

If the two real planes thus determined be $x_4 = \alpha_1 x_3$, $x_4 = \alpha_2 x_3$ where $\alpha_1\alpha_2 = -1$, substituting in the first equation we have four solutions, viz. those given by

$$x_1^2 + x_3^2\,\{p\,(1 - \alpha_1^2) + 2q\alpha_1\} = 0,$$

and by
$$x_1^2 + x_3^2\,\{p\,(1 - \alpha_2^2) + 2q\alpha_2\} = 0,$$

i.e. by
$$x_1^2 - \frac{x_3^2}{\alpha_1^2}\{p\,(1 - \alpha_1^2) + 2q\alpha_1\} = 0.$$

Hence we have *two* real solutions only, i.e. there are only two real tangents to the curve of intersection from the vertex of either cone on which it lies. Hence the curve consists of one portion only. In the case therefore in which only three roots of $F(\lambda) = 0$ are real three focal curves are real and consist in each case of only one portion.

If five roots of $F(\lambda) = 0$ are real, any pair S_i, Q_i have a real self-polar tetrahedron; by the method immediately preceding it can be at once seen that their intersection is either imaginary or consists of two detached portions. Two focal curves are real *.

58. Different forms of the cyclide.

It was seen (Art. 54) that there is always one real pair of surfaces S_i, Q_i consisting of a sphere and an ellipsoid. It will now be shown that if this sphere and ellipsoid have no real intersections

* See Art. 63.

the cyclide consists of two ovals, one within the other. For since the points of the surface are the limiting points of S_i and the tangent planes to Q_i (Art. 56), if S_i lies wholly within Q_i we obtain two sets of points M, M' one within S_i and the other without Q_i, each set forming an oval surface.

When S_i lies wholly without Q_i let σ_1 be the curve along which the transverse common tangent planes of S_i and Q_i touch Q_i, and σ_2 the corresponding curve for direct common tangent planes; then the region between σ_1 and σ_2 gives rise to no real points of the cyclide; the region enclosed by σ_1 gives rise to an oval which cuts S_i orthogonally, the region enclosed by σ_2 gives rise to an oval cutting S_i orthogonally and *enclosing the first oval*, since the tangent planes in the case of σ_2 are more remote from S_i than those for σ_1, so that if a line through O meets the surface in the pairs of points M_1, M_1'; M_2, M_2' it is seen that M_2 is nearer O than either M_1 or M_1', and M_2' is more distant from O than either M_1 or M_1'. Hence one oval encloses the other.

If the focal curve (S_i, Q_i) is real and consists of two portions σ_1, σ_2, the portion of Q_i included within S_i may consist of one connected portion (as in the case of a sphere meeting a spheroid whose axis of revolution is its greater axis), the portions of Q_i giving rise to real points of the cyclide are entirely separated, and it consists of two separated ovals (each meeting S_i orthogonally); or the portion of Q_i within S_i may consist of two separate portions (as in the case of a sphere meeting a spheroid whose axis of revolution is its minor axis); here the portion of Q_i giving rise to real points of the cyclide is one connected region; the cyclide consists of a tubular surface similar to an anchor-ring or *tore*. Finally, if the focal curve (S_i, Q_i) consists of one portion only, we have one oval cutting S_i orthogonally.

59. Equal roots of the fundamental quintic.

If $(\lambda + A_1)^2$ is a root of $F(\lambda)$, then V (Art. 54) is a pair of planes; for if the A_i are all unequal, then we must have

$$B_1 = 0, \quad \begin{vmatrix} A_2 - A_1 & 0 & B_2 \\ 0 & A_3 - A_1 & B_3 \\ B_2 & B_3 & C - A_1^2 \end{vmatrix} = 0,$$

which makes V a pair of planes when $\lambda + A_1 = 0$.

The equation of the surface is $S^2 + \alpha\beta = 0$; inverting from one of the points (S, α, β) we obtain a cone K; three sets of bitangent spheres of the surface are therefore the inverses of spheres passing through a pair of circular sections of K.

Again, if $A_1 = A_2$, then if $(\lambda + A_1)^2$ is a factor of $F(\lambda)$, we must have $B_1^2 + B_2^2 = 0$, i.e. $B_1 = B_2 = 0$ in a real cyclide, and the surface has an equation of the form

$$S^2 + \alpha\alpha' = 0,$$

where α and α' are parallel planes.

The ∞^2 spheres $S + \lambda\alpha + \mu\alpha' = 0$ meet the surface in pairs of circles. These spheres consist of *all spheres having their centres on a given line.* The surface has also three sets of bitangent spheres.

When $A_1 = A_2 = A_3$, the equation of the surface may be written

$$S^2 + k\alpha = 0.$$

The ∞^2 spheres $S = \lambda\alpha + \mu$, which are all spheres having their centres on a given line, meet the surface in pairs of circles.

In each of these cases, therefore, one of the five cones V is a pair of planes*.

In a real cyclide only one of the principal spheres can be a point-sphere. For it has been seen (Art. 55) that if the A_i are unequal there lie an odd number of roots of $F(\lambda) = 0$ in each of the three intervals

$$-\infty \ldots -A_1, \quad -A_1 \ldots -A_2, \quad -A_2 \ldots -A_3.$$

Hence coincidence of roots of $F(\lambda) = 0$ can only occur *once.*

Again, if two of the A_i are equal, say $A_1 = A_2$, then, excluding the case which has been already considered in which $(\lambda + A_1)^2$ is a factor of $F(\lambda)$, we have

$$F(\lambda) = (\lambda + A_1)\,\psi(\lambda),$$

where $\psi(\lambda)$ is seen as before to have an odd number of real roots in each of two intervals. It therefore follows that $F(\lambda)$ may have *one* double root or *one* triple root; in each of these cases the remaining roots are real.

If R_i be the radius of the principal sphere S_i,

$$R_i^2 = -2\lambda_i + \frac{B_1^2}{(A_1 + \lambda_i)^2} + \frac{B_2^2}{(A_2 + \lambda_i)^2} + \frac{B_3^2}{(A_3 + \lambda_i)^2} = \phi'(\lambda_i);$$

* The surface is of the type [(11) 111], see Art. 67.

and since $F(\lambda)$ does not possess $(\lambda + A_i)^2$ as a factor, a principal point-sphere will arise from equal roots of $F(\lambda)$, which, it has been seen, can occur for only one value of λ.

In all cases, therefore, a real cyclide can have only *one* principal point-sphere; the case in which one of the cones V is a pair of planes will be discussed later (Arts. 66, 67).

60. Power of two spheres.

If $S_1 = 0, \ldots S_5 = 0$ are any five spheres, the system of five equations

$$x^2 + y^2 + z^2 + 2f_1xw + 2g_1yw + 2h_1zw + c_1w^2 \equiv \rho S_1,$$
$$\cdots\cdots\cdots\cdots\cdots\cdots\cdots\cdots\cdots\cdots\cdots\cdots$$
$$x^2 + y^2 + z^2 + 2f_5xw + 2g_5yw + 2h_5zw + c_5w^2 \equiv \rho S_5,$$

wherein $w = 1$, enables us to solve for $x^2 + y^2 + z^2$, xw, yw, zw, w^2 in terms of $S_1 \ldots S_5$; this gives rise to a quadratic identity between the quantities $S_1 \ldots S_5$. These five quantities may be employed as coordinates to determine the position of a point, a homogeneous quadratic relationship existing between the coordinates. These coordinates are known as the *pentaspherical coordinates* of a point. The nature of this quadratic relationship can be most readily determined from considerations relating to the mutual *power* of two spheres. If two spheres of radii r_1, r_2 cut one another at an angle θ, we have

$$\begin{aligned} 2r_1r_2\cos\theta &= r_1^2 + r_2^2 - (f_1 - f_2)^2 - (g_1 - g_2)^2 - (h_1 - h_2)^2 \\ &= 2f_1f_2 + 2g_1g_2 + 2h_1h_2 - c_1 - c_2. \end{aligned}$$

The right-hand side of these equations is real for real spheres whether their intersection be real or otherwise; taken negatively it is known as the mutual *power*, π_{12}, of the two spheres, thus

$$\pi_{12} = c_1 + c_2 - 2f_1f_2 - 2g_1g_2 - 2h_1h_2.$$

Forming the product of the two determinants*

$$\begin{vmatrix} 0 & 1 & 2f_1 & 2g_1 & 2h_1 & c_1 \\ 0 & 1 & 2f_2 & 2g_2 & 2h_2 & c_2 \\ 0 & 1 & 2f_3 & 2g_3 & 2h_3 & c_3 \\ 0 & 1 & 2f_4 & 2g_4 & 2h_4 & c_4 \\ 0 & 1 & 2f_5 & 2g_5 & 2h_5 & c_5 \\ 0 & 1 & 2f_6 & 2g_6 & 2h_6 & c_6 \end{vmatrix}, \quad \begin{vmatrix} 0 & c_7 & -f_7 & -g_7 & -h_7 & 1 \\ 0 & c_8 & -f_8 & -g_8 & -h_8 & 1 \\ 0 & c_9 & -f_9 & -g_9 & -h_9 & 1 \\ 0 & c_{10} & -f_{10} & -g_{10} & -h_{10} & 1 \\ 0 & c_{11} & -f_{11} & -g_{11} & -h_{11} & 1 \\ 0 & c_{12} & -f_{12} & -g_{12} & -h_{12} & 1 \end{vmatrix},$$

* Lachlan, *On systems of circles and spheres*, Roy. Soc. Trans. (1886).

we obtain

$$\begin{vmatrix} \pi_{1,7} & \pi_{1,8} & \pi_{1,9} & \pi_{1,10} & \pi_{1,11} & \pi_{1,12} \\ \pi_{2,7} & \pi_{2,8} & \pi_{2,9} & \pi_{2,10} & \pi_{2,11} & \pi_{2,12} \\ \pi_{3,7} & \pi_{3,8} & . & . & . & . \\ \vdots & & & & & \\ \pi_{6,7} & \pi_{6,8} & . & . & . & \pi_{6,12} \end{vmatrix} = 0.$$

This equation may be denoted by $\pi\begin{pmatrix}1 \dots & 6 \\ 7 \dots & 12\end{pmatrix} = 0.$

Now, denoting the spheres S_1 and S_7 by x and y, and supposing that the spheres $S_2 \dots S_6$ are respectively identical with the spheres $S_8 \dots S_{12}$, we have on slightly altering the notation

$$\pi\begin{pmatrix} x & 1 \dots 5 \\ y & 1 \dots 5\end{pmatrix} = 0,$$

which, expanded, is equivalent to

$$\begin{vmatrix} \pi_{xy} & \pi_{x1} \dots \pi_{x5} \\ \pi_{y1} & \pi_{11} \dots \pi_{15} \\ \vdots & \\ \pi_{y5} & \pi_{15} \dots \pi_{55} \end{vmatrix} = 0 \;\dots\dots\dots\dots\dots\dots(1).$$

If we now suppose $S_x \equiv S_y$, we obtain the relationship existing between the powers of any sphere S_x with regard to five fixed spheres, viz.

$$\begin{vmatrix} -2r^2 & \pi_{x1} & . & . & . & \pi_{x5} \\ \pi_{x1} & -2r_1^2 & \pi_{12} & . & . & . \\ . & \pi_{12} & -2r_2^2 & . & . & . \\ . & . & . & -2r_3^2 & . & . \\ . & . & . & . & -2r_4^2 & . \\ \pi_{x5} & . & . & . & . & -2r_5^2 \end{vmatrix} = 0 \;\dots(2).$$

If this equation is such that π_{x1} occurs only in the term involving π_{x1}^2, the sphere S_1 cuts $S_2 \dots S_5$ orthogonally; for the coefficients of $\pi_{x1}\pi_{x2}, \dots \pi_{x1}\pi_{x5}$ all involve $\pi_{12}, \dots \pi_{15}$ linearly and homogeneously, hence if they all vanish we have either

$$\pi_{12} = \pi_{13} = \pi_{14} = \pi_{15} = 0,$$

or

$$\begin{vmatrix} \pi_{22} & \pi_{23} & \pi_{24} & \pi_{25} \\ \pi_{23} & . & . & . \\ \pi_{24} & . & . & . \\ \pi_{25} & . & . & . \end{vmatrix} = 0.$$

But the last condition cannot be fulfilled since the determinant on the left side is equal to the coefficient of π_{x1}^2, taken negatively, and this is by hypothesis not zero.

If the sphere S_x is a point-sphere, its powers with respect to the spheres $S_1 \dots S_5$ are obtained by substituting the coordinates of its centre (x, y, z) in the expressions $S_1 \dots S_5$; hence we obtain the required identical relation between any five spheres $S_1 = 0, \dots S_5 = 0$, which is therefore

$$\begin{vmatrix} 0 & S_1 & \cdot & \cdot & \cdot & S_5 \\ S_1 & -2r_1^2 & \pi_{12} & \cdot & \cdot & \pi_{15} \\ \cdot & \pi_{12} & -2r_2^2 & \cdot & \cdot & \cdot \\ \cdot & \cdot & \cdot & -2r_3^2 & \cdot & \cdot \\ \cdot & \cdot & \cdot & \cdot & -2r_4^2 & \cdot \\ S_5 & \cdot & \cdot & \cdot & \cdot & -2r_5^2 \end{vmatrix} = 0.$$

It follows, as in the case just above, that if S_i occurs only in the form S_i^2, the sphere S_i cuts orthogonally the remaining four spheres. If all the quantities $\pi_{ij} (i \neq j)$ vanish, the identical relation becomes

$$\sum_1^5 \left(\frac{S_i}{r_i}\right)^2 = 0,$$

and the five spheres are mutually orthogonal.

By virtue of this equation, the equation of the sphere S_1 (say) may be written in the form

$$\left(\frac{S_1 - S_2}{r_2}\right)^2 + \left(\frac{S_1 - S_3}{r_3}\right)^2 + \left(\frac{S_1 - S_4}{r_4}\right)^2 + \left(\frac{S_1 - S_5}{r_5}\right)^2 = 0.$$

This shows that the planes of intersection of S_1 with S_2, S_3, S_4 and S_5 form a self-polar tetrahedron for S_1. Now the radical plane of S_1 and S_2 contains the centres of S_3, S_4 and S_5, and so on; hence we again obtain the result of Art. 55 that the centres of any four spheres form a self-polar tetrahedron for the fifth sphere.

We observe that if four of the spheres S_i, supposed mutually orthogonal, are real, the fifth sphere is also real but the square of its radius is negative.

Also we see that on inverting from any point not upon one of the spheres S_i, the form of the relationship is not altered, since

$$\frac{S_i}{r_i} \propto \frac{S_i'}{r_i'}.$$

But if the centre of inversion be a point of intersection of three spheres S_1, S_2, S_3, they are inverted into three planes which we may take to be coordinate planes, and since $\frac{S_i}{r_i} \propto 2k^2x$, etc., we obtain the identity

$$4k^4(x^2+y^2+z^2)+\frac{(x^2+y^2+z^2-R^2)^2}{R^2}k^4-\frac{(x^2+y^2+z^2+R^2)^2}{R^2}k^4=0;$$

where R^2 and $-R^2$ are the squares of the radii of the spheres into which S_4 and S_5 are inverted.

When the identical relation has the form

$$AS_1S_2+BS_3^2+CS_4^2+DS_5^2=0,$$

it follows from the preceding case that $r_1=r_2=0$, hence S_1 and S_2 are point-spheres whose centres are the intersections of the spheres S_3, S_4 and S_5.

When the relation is

$$AS_1S_2+BS_3S_4+CS_5^2=0;$$

S_1 and S_2 are point-spheres, and also S_3 and S_4. The centres of one pair of point-spheres lie on the intersections of the other pair; hence one pair is real, the other is conjugate imaginary; the centres of all four point-spheres lie on S_5.

61. Sphere referred to five orthogonal spheres.

The equation of any sphere S may be expressed in terms of any five mutually orthogonal spheres, thus if

$$S \equiv x^2+y^2+z^2+2fx+2gy+c=0,$$

and also if $S \equiv \sum_1^5 a_iS_i$; then, denoting by π_{a_i} the power of S with regard to the sphere S_i, we have

$$-\pi_{a_i}=2rr_i\cos\theta_i=\frac{2f_i\Sigma af+2g_i\Sigma ag+2h_i\Sigma ah}{\Sigma a}-c_i-\frac{\Sigma ac}{\Sigma a};$$

hence, from the fact that $S_1 \dots S_5$ are mutually orthogonal

$$-\pi_{a_i}=\frac{2a_i}{\Sigma a}(f_i^2+g_i^2+h_i^2-c_i)=\frac{2a_ir_i^2}{\Sigma a}.$$

Hence $\qquad \pi_{a_i}=\rho a_ir_i^2, \quad \cos\theta_i=\sigma a_ir_i.$

Introducing into equation (1) the angles θ_i, ϕ_i at which two spheres intersect any five spheres $S_1 \dots S_5$, then if ψ be the angle at which the two spheres intersect, we obtain

$$\begin{vmatrix} \cos\psi & \cos\theta_1 & \dots\dots\dots & \cos\theta_5 \\ \cos\phi_1 & 1 & \cos\widehat{12} & \dots\dots \\ \vdots & & & \\ \cos\phi_5 & \dots\dots & \dots\dots & 1 \end{vmatrix}=0.$$

If the spheres $S_1 \dots S_5$ are mutually orthogonal this reduces to

$$\cos\psi = \sum_1^5 \cos\theta_i \cos\phi_i.$$

If the two spheres are identical we obtain $\sum_1^5 \cos^2\theta_i = 1$; if they are $\sum_1^5 a_i S_i = 0$, $\sum_1^5 b_i S_i = 0$, and cut orthogonally we have from above $\sum_1^5 a_i b_i r_i^2 = 0$.

If S is orthogonal to one sphere of the orthogonal system $S_1 \dots S_5$, say to S_5, the equation of S is $\sum_1^4 a_i S_i = 0$. In this case the volume of the tetrahedron whose base is the triangle formed by the centres of S_2, S_3, S_4 and whose vertex is the centre of S_1 is

$$-\frac{1}{\Sigma a}\begin{vmatrix} \sum_1^4 a_i f_i & \sum_1^4 a_i g_i & \sum_1^4 a_i h_i & \sum_1^4 a_i \\ f_2 & g_2 & h_2 & 1 \\ f_3 & g_3 & h_3 & 1 \\ f_4 & g_4 & h_4 & 1 \end{vmatrix} = -\frac{a_1}{\sum_1^4 a_i}\begin{vmatrix} f_1 & g_1 & h_1 & 1 \\ f_2 & g_2 & h_2 & 1 \\ f_3 & g_3 & h_3 & 1 \\ f_4 & g_4 & h_4 & 1 \end{vmatrix}.$$

Hence if $\xi_1 \dots \xi_4$ are the tetrahedral coordinates of the centre of S with regard to the tetrahedron formed by the centres of $S_1 \dots S_4$, we have that $\xi_i \propto a_i$.

When the sphere $\sum_1^5 a_i S_i = 0$ is a point-sphere, we have $\sum_1^5 \frac{\pi_{a_i}^2}{r_i^2} = 4r^2 \sum_1^5 \cos^2\theta_i$, which is zero since $\Sigma \cos^2\theta_i = 1$ and $r = 0$. In this case $\sum_1^5 a_i^2 r_i^2 = 0$, so that if $x_i = \frac{S_i}{r_i}$, the equation of a point-sphere is $\sum_1^5 a_i x_i = 0$, with the condition $\Sigma a_i^2 = 0$.

62. Pentaspherical coordinates.

It has been seen (Art. 60) that the quantities $x^2 + y^2 + z^2$, x, y, z and unity which occur in the equation of any cyclide may be replaced by linear functions of $S_1 \dots S_5$; the equation of the cyclide then appears as a quadratic in the S_i which are themselves connected by a quadratic identity. The quantities $\frac{S_i}{r_i}$, or x_i, are termed the *pentaspherical coordinates* of a point.

If the equation of the surface expressed in pentaspherical coordinates contains only four of the variables, say $x_1 \dots x_4$, so that its equation is $\sum_1^4 a_{ik} x_i x_k = 0$, the surface is clearly the envelope of the sphere $\sum_1^4 a_i x_i = 0$, where the coefficients a_i are subject to the condition $\sum_1^4 A_{ik}\alpha_i\alpha_k = 0$. Since by the last article the α_i are the coordinates of the centre of this sphere, we obtain the cyclide as the envelope of a sphere which cuts a fixed sphere orthogonally and whose centre lies on a quadric.

63. Canonical forms of the equation of the cyclide.

The equation of the cyclide being quadratic in five variables $x_1 \ldots x_5$ which are themselves connected by an identical equation; we may use the method of Elementary Factors* to obtain the various types of canonical forms of the cyclide.

Denoting by $\Phi = 0$ the equation of the cyclide and by $\Omega = 0$ the identical relation connecting the coordinates, we obtain by this method seven types, viz.

[11111], [2111], [311], [221], [41], [32], [5];

each type giving rise to sub-types. It will also be seen that only the first three forms relate to *real* cyclides.

Writing these forms at length, we obtain by the usual method

$$[11111] \begin{cases} \Phi \equiv \lambda_1 x_1^2 + \lambda_2 x_2^2 + \lambda_3 x_3^2 + \lambda_4 x_4^2 + \lambda_5 x_5^2, \\ \Omega \equiv x_1^2 + x_2^2 + x_3^2 + x_4^2 + x_5^2. \end{cases}$$

$$[2111] \begin{cases} \Phi \equiv 2\lambda_1 x_1 x_2 + x_1^2 + \lambda_3 x_3^2 + \lambda_4 x_4^2 + \lambda_5 x_5^2, \\ \Omega \equiv 2x_1 x_2 + x_3^2 + x_4^2 + x_5^2. \end{cases}$$

$$[311] \begin{cases} \Phi \equiv \lambda_1 (2x_1 x_3 + x_2^2) + 2x_1 x_2 + \lambda_4 x_4^2 + \lambda_5 x_5^2, \\ \Omega \equiv 2x_1 x_3 + x_2^2 + x_4^2 + x_5^2. \end{cases}$$

$$[221] \begin{cases} \Phi \equiv 2\lambda_1 x_1 x_2 + x_1^2 + 2\lambda_3 x_3 x_4 + x_3^2 + \lambda_5 x_5^2, \\ \Omega \equiv 2x_1 x_2 + 2x_3 x_4 + x_5^2. \end{cases}$$

$$[41] \begin{cases} \Phi \equiv 2\lambda_1 (x_1 x_4 + x_2 x_3) + 2x_1 x_3 + x_2^2 + \lambda_5 x_5^2, \\ \Omega \equiv 2(x_1 x_4 + x_2 x_3) + x_5^2. \end{cases}$$

$$[32] \begin{cases} \Phi \equiv \lambda_1 (2x_1 x_3 + x_2^2) + 2x_1 x_2 + 2\lambda_4 x_4 x_5 + x_4^2, \\ \Omega \equiv 2x_1 x_3 + x_2^2 + 2x_4 x_5. \end{cases}$$

$$[5] \begin{cases} \Phi \equiv \lambda (2x_1 x_5 + 2x_2 x_4 + x_3^2) + 2x_1 x_4 + 2x_2 x_3, \\ \Omega \equiv 2x_1 x_5 + 2x_2 x_4 + x_3^2. \end{cases}$$

We now pass to consideration of the type [11111]; the form of Ω shows that here the coordinate spheres form an orthogonal

* For discussion of the method see Bromwich, *Quadratic forms and their classification by means of invariant factors* (Cambridge Tracts), or the Author's *Treatise on the Line Complex.* See also Bôcher's *Potential Theorie.*

system: eliminating one of the variables, say x_5, we obtain an equation of the form

$$\sum_1^4 (\lambda_i - \lambda_5)\, x_i^2 = 0 \quad \text{.....................(1).}$$

The surface is therefore the envelope of the spheres

$$\sum_1^4 \alpha_i x_i = 0$$

subject to the condition

$$\sum_1^4 \frac{\alpha_i^2}{\lambda_i - \lambda_5} = 0 \quad \text{.......................(2).}$$

The generating sphere is orthogonal to $x_5 = 0$; and since the α_i are the coordinates of its centre for the tetrahedron formed by the centres of $S_1 \dots S_4$ (Art. 61), the equation (2) represents the quadric Q_5.

We obtain similarly the other four sets of generating bitangent spheres. Moreover, assuming the cyclide to be real, and since it was seen (Art. 54) that the only bitangent spheres are those arising from a pair S_i, Q_i of this cyclide, it follows that the spheres $x_i = 0$ can be no other than the spheres S_i which a real cyclide possesses. Hence it follows that of these spheres at least three are real, while two may be either real or conjugate imaginary; so that this applies to the variables x_i; and if e.g. x_4 and x_5 are conjugate imaginary, so also are λ_4 and λ_5; if all the x_i are real, so also are the λ_i.

The five focal curves are determined by the equations

$$\sum_1^4 \frac{\alpha_i^2}{\lambda_i - \lambda_5} = 0, \quad \sum_1^4 \alpha_i^2 = 0,$$

together with four other similar pairs of equations.

It was seen (Art. 57) that if two of the principal spheres are conjugate imaginary three of the focal curves are real and consist of one portion. Consider now the case in which all the principal spheres x_i are real and let $x_5 = 0$ be that sphere the square of whose radius is negative, so that $a_1 \dots a_4$ are real and $a_5 = a_5 r_5$ is imaginary. One of the focal curves is then given by

$$\frac{a_2^2}{\lambda_2 - \lambda_1} + \frac{a_3^2}{\lambda_3 - \lambda_1} + \frac{a_4^2}{\lambda_4 - \lambda_1} + \frac{a_5^2}{\lambda_5 - \lambda_1} = 0,$$

$$\sum_2^5 a_i^2 = 0.$$

This curve is then real or imaginary according as the cone

$$a_2^2 \frac{\lambda_2 - \lambda_5}{\lambda_1 - \lambda_2} + a_3^2 \frac{\lambda_3 - \lambda_5}{\lambda_1 - \lambda_3} + a_4^2 \frac{\lambda_4 - \lambda_5}{\lambda_1 - \lambda_4} = 0$$

does or does not contain real points apart from its vertex.

We obtain in this way criteria for the reality or otherwise of four focal curves; the fifth, which lies upon

$$\sum_1^4 a_i^2 = 0,$$

is of course imaginary.

To discriminate in the four cases we may suppose the quantities $\lambda_1 \ldots \lambda_4$ to be in algebraic order of magnitude and moreover we have as a condition of reality of the surface that the quantities

$$\lambda_1 - \lambda_5, \quad \lambda_2 - \lambda_5, \quad \lambda_3 - \lambda_5, \quad \lambda_4 - \lambda_5$$

cannot all have the same sign. We may take λ_5 equal to unity, in which case $\lambda_1 - \lambda_5$ must be positive, and then the three possible distributions of signs to $\lambda_2 - 1$, $\lambda_3 - 1$, $\lambda_4 - 1$ are

$$+ + -, \quad + - -, \quad - - -.$$

Inserting these signs in the equations of the four cones obtained as above it is seen that in all cases two of them are real and two are imaginary.

64. Form of the cyclide.

In the case in which the variables are all real, x_5 being that principal sphere the square of whose radius is negative, the equation of the surface is

$$\sum_1^4 (\lambda_i - \lambda_5) x_i^2 = 0.$$

If we invert the surface from one point of intersection of the spheres x_1, x_2, x_3 and take as new coordinate planes those into which these spheres are inverted, the equation of the new surface is (Art. 60)

$$4(\lambda_1 - \lambda_5) x^2 + 4(\lambda_2 - \lambda_5) y^2 + 4(\lambda_3 - \lambda_5) z^2 + \frac{(\lambda_4 - \lambda_5)}{R^2}(x^2 + y^2 + z^2 - R^2)^2 = 0,$$

or

$$V + (\lambda_4 - \lambda_5) S^2 = 0.$$

We may assume $\lambda_4 - \lambda_5$ to be positive; different forms of the surface will then arise according as *one, two,* or *three* of the remaining coefficients are negative.

If one of them is negative then every line through the origin and *within* the cone V meets the surface in four real points; hence the surface consists of two ovals, one within each portion of V. If two of them are negative, then every line through the origin and *without* the cone V meets the surface in four real points; the surface is ring-shaped. If all are negative, then every line through the origin meets the surface in four real points; hence the surface consists of two ovals, one within the other, and each surrounding the origin.

Hence the form of the original surface is also determined in this manner by the signs of the quantities $\lambda_i - \lambda_5$. When all the variables are real, the inverted surface is seen to be derived from the general cyclide by taking B_1, B_2 and B_3 all zero and C positive (Art. 54). This is one form of cyclide with *three planes of symmetry*.

The other form in which C is negative corresponds to the case when two of the variables are imaginary; for in this case we have

$$x_1^2 + x_2^2 + \sum_3^5 x_i^2 = 0,$$

in which we may take

$$x_1 = \xi_1 + i\xi_2, \quad x_2 = \xi_1 - i\xi_2;$$

substituting these values for x_1 and x_2 the identical relation assumes the form

$$2(\xi_1 + \xi_2)(\xi_1 - \xi_2) + \sum_3^5 x_i^2 = 0.$$

Hence $\xi_1 + \xi_2$ and $\xi_1 - \xi_2$ are point-spheres.

If we make the same substitution in the equation of the cyclide, it assumes the form (in which only real quantities occur),

$$A(\xi_1^2 - \xi_2^2) + 4B\xi_1\xi_2 + \sum_3^5 \lambda_i x_i^2 = 0,$$

that is

$$A(\xi_1^2 - \xi_2^2) + B\{(\xi_1 + \xi_2)^2 - (\xi_1 - \xi_2)^2\} + \sum_3^5 \lambda_i x_i^2 = 0.$$

Inverting from an intersection of the spheres x_3, x_4, x_5, i.e. from the centre of one of the point-spheres $\xi_1 + \xi_2$, $\xi_1 - \xi_2$, the equation of the inverse surface is seen to be of the form

$$(x^2 + y^2 + z^2)^2 + (\lambda_3 + \kappa)x^2 + (\lambda_4 + \kappa)y^2 + (\lambda_5 + \kappa)z^2 - m^2 = 0.$$

In this case every line through the origin meets the surface in two real and in two imaginary points.

65. The type [2111].

The equations determining the second type show that it represents a cyclide which can be generated in four ways; viz. in three ways by bitangent spheres orthogonal to three given spheres respectively, and once by a sphere passing through a given point, which is one of the intersections of the spheres x_3, x_4 and x_5. Two of the principal spheres, S_1 and S_2 of the general case, here come into coincidence with the point-sphere x_1. It follows from Art. 59 that the principal spheres are all real.

That this is a degenerate case of the general case may be seen as follows: let us change the notation and write

$$\Omega \equiv \sum_1^5 a_i x_i^2, \quad \Phi \equiv \sum_1^5 \lambda_i a_i x_i^2,$$

and let $\lambda_2 = \lambda_1 + \epsilon$, $x_2 = x_1 + \epsilon x_2'$, where ϵ is small.

Then $$\Omega \equiv (a_1 + a_2) x_1^2 + 2a_2 \epsilon x_1 x_2' + \sum_3^5 a_i x_i^2,$$

$$\Phi = \lambda_1 (a_1 + a_2) x_1^2 + a_2 \epsilon (x_1^2 + 2\lambda_1 x_1 x_2') + \sum_3^5 \lambda_i a_i x_i^2.$$

If we now assume that

$$a_1 + a_2 = 0, \quad a_2 \epsilon = 1, \quad a_3 = a_4 = a_5 = 1,$$

we obtain the second type. See Bôcher, *Potential Theorie.*

The surface has a node, the centre of the point-sphere x_1.

If we invert the surface from this node, we obtain the quadric

$$(\lambda_3 - \lambda_1) x^2 + (\lambda_4 - \lambda_1) y^2 + (\lambda_5 - \lambda_1) z^2 + h^2 = 0.$$

Hence, if the node is isolated the surface is the inverse of an ellipsoid; otherwise it will be ring-shaped if it is the inverse of a hyperboloid of one sheet; it will consist of two sheets united at the node if it is the inverse of a hyperboloid of two sheets.

That the cyclide is the inverse of a quadric when one of the principal spheres reduces to a point, may also be seen as follows: if Q is the quadric associated with the point-sphere O, the surface is the envelope of spheres passing through O and having their centres on Q; all the spheres whose centres are consecutive to any point P of Q will pass through the point O' which is the image of O for the tangent plane of Q at P. Hence the surface is similar to the pedal surface of Q for O, and is therefore similar to the inverse of the reciprocal polar of Q for O.

66. The type [311].

The equations connected with this type of cyclide show that it is generated in three ways; twice as the envelope of a sphere cutting orthogonally two given spheres respectively; and once as the envelope of a sphere passing through a given point; the spheres x_2, x_3 of the general case come into coincidence with the point-sphere x_1.

The equation of the surface being

$$(\lambda_4 - \lambda_1) x_4^2 + (\lambda_5 - \lambda_1) x_5^2 + 2x_1 x_2 = 0,$$

the centre of the point-sphere x_1 is seen to be a node, and since the spheres x_2, x_4, x_5 pass through this point and cut orthogonally, the tangent cone of the cyclide at the point consists of two planes.

Inverting the surface from the node, we obtain as the inverse surface, the paraboloid

$$(\lambda_2 - \lambda_1) x^2 + (\lambda_3 - \lambda_1) y^2 + 2kz = 0;$$

which is elliptic or hyperbolic according as the biplanar node has imaginary or real tangent planes.

The four remaining types give rise to cyclides which are imaginary; for the quintic $F(\lambda) = 0$ may have either one double root or one triple root, but no other coincidence of roots can occur in a real cyclide (Art. 59)*.

67. The sub-type [(11) 111].

The sub-types arising from the above three chief types, as for instance [(11) 111], are such that the equation of the surface can be expressed in terms of only three variables; thus [(11) 111] has an equation of the form

$$Ax_3^2 + Bx_4^2 + Cx_5^2 = 0.$$

The common characteristic of all the sub-types is that one of the five cones V should be a pair of planes, real or imaginary. For, if in the equation

$$(a \between S_1, S_2, S_3)^2 = 0,$$

we substitute $S_2 = a_2 S_1 + \alpha$, $S_3 = a_3 S_1 + \beta$, where α and β are linear in the coordinates, we obtain an equation of the form

$$(S_1 + \gamma)^2 = (b \between \alpha, \beta)^2;$$

γ being linear in the variables.

The surface [(11) 111] has two nodes which may be either real or imaginary. If the nodes are real, on inverting the surface from one of them we obtain a quadric cone.

The cyclide $Ax_3^2 + Bx_4^2 + Cx_5^2 = 0$ is the envelope of the spheres $\alpha_3 x_3 + \alpha_4 x_4 + \alpha_5 x_5 = 0$, subject to the condition

$$\frac{\alpha_3^2}{A} + \frac{\alpha_4^2}{B} + \frac{\alpha_5^2}{C} = 0 \quad \text{.......................(1).}$$

The contact of these spheres and their envelope occurs along a circle instead of at two points.

* The cyclide $S^2 = \alpha\beta$ (Art. 59) is always expressible in the form $(S_1, S_2, S_3 \between a)^2 = 0$, where S_1, S_2, S_3 are three mutually orthogonal spheres; hence it cannot belong to one of the types [221], [41], [32], [5].

These spheres cut both x_1 and x_2 orthogonally, hence they pass through the two limiting points of x_1, x_2, so that their centres lie in a plane; since they also lie on the cone (1) they lie on a conic. This surface is therefore *the envelope of a sphere which passes through a fixed point and whose centre lies on a conic.*

Two systems of bitangent spheres coincide with this system, the other three, which may be called the *proper* systems, remaining as before. They are obtained by writing the equation of the cyclide in the form

$$(B-A)\,x_4^2+(C-A)\,x_5^2-A\,(x_1^2+x_2^2)=0,$$

showing that the surface is the envelope of the spheres

$$\alpha_1 x_1+\alpha_2 x_2+\alpha_4 x_4+\alpha_5 x_5=0,$$

subject to the condition

$$\frac{\alpha_1^2}{A}+\frac{\alpha_2^2}{A}-\frac{\alpha_4^2}{B-A}-\frac{\alpha_5^2}{C-A}=0 \quad \text{..............(2).}$$

Now in the preceding equation (1) we may assume

$$\alpha_3^2+\alpha_4^2+\alpha_5^2=1,$$

hence it is equivalent to

$$\alpha_4^2\,\frac{B-A}{B}+\alpha_5^2\,\frac{C-A}{C}=1 \quad \text{..................(3).}$$

Also in equation (2) we may assume that

$$\alpha_1^2+\alpha_2^2+\alpha_4^2+\alpha_5^2=1\,;$$

hence it is equivalent to

$$\alpha_4^2\,\frac{B}{B-A}+\alpha_5^2\,\frac{C}{C-A}=1 \quad \text{..................(4).}$$

These equations (3) and (4) hold respectively for generating spheres of the special system and a proper system.

Now take two fixed generating spheres of the special system whose coordinates are $(0, 0, v_1, z_1, w_1)$, $(0, 0, v_2, z_2, w_2)$, and a variable sphere of the system to which (4) relates whose coordinates are $(x, y, 0, z, w)$; then if ϕ_1, ϕ_2 respectively are the angles at which the variable sphere cuts the fixed spheres, we have (Art. 61)

$$\cos\phi_1 = zz_1+ww_1$$
$$=z\sqrt{\frac{B}{B-A}}\,z_1\sqrt{\frac{B-A}{B}}+w\sqrt{\frac{C}{C-A}}\,w_1\sqrt{\frac{C-A}{C}},$$

$$\cos\phi_2 = zz_2+ww_2$$
$$=z\sqrt{\frac{B}{B-A}}\,z_2\sqrt{\frac{B-A}{B}}+w\sqrt{\frac{C}{C-A}}\,w_2\sqrt{\frac{C-A}{C}}.$$

Hence by virtue of equations (3) and (4) we may take ϕ_1 and ϕ_2 to be the angles which a variable line makes with two fixed lines in its plane; hence

$$\phi_1 \pm \phi_2 = \text{constant}.$$

Therefore the sum or difference of the angles which a variable sphere of one of the three proper systems makes with two fixed spheres of the special system is a constant.

The corresponding result for the general cyclide is the following: the angles which any generating sphere of one set makes with three fixed generating spheres of another set, are equal to the angles which a variable line makes with three fixed lines.

68. Dupin's cyclide.

The surface [(11)(11)1] is known as *Dupin's cyclide*; its equation takes either of the forms

$$(\lambda_1 - \lambda_3)(x_1^2 + x_2^2) + (\lambda_5 - \lambda_3)x_5^2 = 0,$$
$$(\lambda_3 - \lambda_1)(x_3^2 + x_4^2) + (\lambda_5 - \lambda_1)x_5^2 = 0.$$

It has four nodes, of which at least two are conjugate imaginary, since at least two lie upon that sphere the square of whose radius is negative.

Inverting from a node (supposed real) we obtain a cone of revolution.

The spheres which touch the surface along circles form two systems, one of the systems is given by the equations

$$\sum_3^5 \alpha_i x_i = 0, \quad \sum_3^5 \alpha_i^2 = 1, \quad \frac{\alpha_3^2 + \alpha_4^2}{\lambda_3 - \lambda_1} + \frac{\alpha_5^2}{\lambda_5 - \lambda_1} = 0;$$

hence α_5 is constant and equal to $\sqrt{\dfrac{\lambda_5 - \lambda_1}{\lambda_5 - \lambda_3}}$, so that these spheres cut the sphere x_5 at a constant angle. If they lie within x_5, α_5 is positive and greater than unity, say equal to $\sec\beta$; the spheres will therefore touch each of the fixed spheres $(\sin\beta, 0, 0, 0, \cos\beta)$, $(0, \sin\beta, 0, 0, \cos\beta)$. Hence Dupin's cyclide is the envelope of spheres having their centres on a fixed plane and touching each of two fixed spheres. The fixed spheres are not unique, since they are any two of the singly infinite set $(A_1, A_2, 0, 0, \cos\beta)$ where $A_1^2 + A_2^2 = \sin^2\beta$; whose centres lie in the radical plane of x_3 and x_4.

We obtain the same cyclide as the envelope of spheres cutting x_3 and x_4 orthogonally; they have their centres on a second conic whose plane is perpendicular to the line joining the centres of x_3 and x_4. Since the join of the centres of x_1 and x_2 is perpendicular to the join of the centres of x_3 and x_4 these two conics lie in perpendicular planes. These spheres form the previous set $(A_1, A_2, 0, 0, \cos\beta)$; for they are $\alpha_1 x_1 + \alpha_2 x_2 + \alpha_5 x_5 = 0$, with the condition

$$\alpha_1^2 + \alpha_2^2 = \alpha_5^2\left(\frac{\lambda_5 - \lambda_1}{\lambda_5 - \lambda_3} - 1\right) = \alpha_5^2 \tan^2\beta.$$

This cyclide was originally defined as the envelope of a sphere touching three fixed spheres, but such spheres form four distinct sets, each set enveloping a cyclide.

The equation of the tore or anchor-ring is

$$\{x^2 + y^2 + z^2 + c^2 - a^2\}^2 = 4c^2(x^2 + y^2),$$

where c is the distance of the centre of the revolving circle from the axis of revolution, and a its radius; inverting from any point we obtain

$$x_5^2 + A(x_3^2 + x_4^2) = 0.$$

If c is greater than a, then x_5 is a sphere the square of whose radius is negative and the cyclide is a Dupin's cyclide with no real nodes; if c is less than a we then obtain a Dupin's cyclide two of whose nodes are real.

69. We add a list of the various distinct real types of cyclide; the remaining forms consist merely of pairs of spheres, etc.

		Inverse surface	Nodes
[11111]	General cyclide:		
	Surface either has two sheets or is ring-shaped	Cyclide with three planes of symmetry (constant term positive)	
	Surface has one sheet	Cyclide with three planes of symmetry (constant term negative)	
[(11) 111]		Cone	2
[(11) (11) 1]		Cone of revolution	4 (two imaginary)
[2111]		Ellipsoid	1 (isolated)
		Hyperboloid of one sheet	1
		Hyperboloid of two sheets	1
[2 (11) 1]		Ellipsoid or hyperboloid of revolution	3 (two imaginary)

	Inverse surface	Nodes
[(21) 11]	Elliptic or hyperbolic cylinder	1
[(21) (11)]	Circular cylinder	3 (two imaginary)
[311]	Paraboloid	1 (biplanar)
[3 (11)]	Paraboloid of revolution	3 (two imaginary)
[(31) 1]	Parabolic cylinder	1 (uniplanar)

70. Tangent spheres of the cyclide.

The equation $\sum\limits_1^5 (a_i+\lambda)\, x_i y_i = 0$, where the x_i are the coordinates of a point of the cyclide $\sum\limits_1^5 a_i x_i^2 = 0$, represents the ∞^1 tangent spheres of the cyclide at the point x_i.

Hence* if $\Sigma\, m_i y_i = 0$ is a tangent sphere of the cyclide, we have

$$\Sigma\, \frac{m_i^2}{a_i+\lambda} = 0, \quad \Sigma\, \frac{m_i^2}{(a_i+\lambda)^2} = 0.$$

Eliminating λ between these equations, we have the relation fulfilled by the coordinates m_i of a tangent sphere of the cyclide. It arises by expressing that the equation $\Sigma\, \dfrac{m_i^2}{a_i+\lambda} = 0$ should have a double root. The equation being of the fourth degree in λ its discriminant will be of the twelfth order in the m_i.

Let $\psi(m) = 0$ represent this equation, $m_i + \rho m_i'$ represents the ∞^1 spheres passing through the intersection of any two spheres m_i and m_i', and we obtain those spheres which touch the cyclide by means of the equation

$$\psi(m + \rho m') = 0,$$

which is of the twelfth degree in ρ. Hence *through any given circle twelve spheres may be drawn to touch the cyclide.*

If the spheres m and m' are concentric,

$$m_i' = \kappa m_i + \frac{\sigma}{r_i}\dagger,$$

and the equation $\psi(m + \rho m') = 0$ gives the twelve spheres having the given point as centre, which can be drawn to touch the cyclide, i.e. *twelve normals can be drawn from any given point to a cyclide.*

* See Darboux, *Sur une classe remarquable*, etc., p. 275.

† Since $\Sigma\, \dfrac{S_i}{r_i^2} \equiv$ constant.

Bitangent spheres.

If in the spheres $\sum_1^5 (a_i+\lambda)x_iy_i=0$ we take λ to be successively $-a_1 \ldots -a_5$, we obtain the five bitangent spheres which touch the cyclide at the point x; e.g. if $\lambda+a_1=0$, the corresponding sphere touches the cyclide in the two points $\pm x_1, x_2 \ldots x_5$.

The Cartesian coordinates of the centre of the sphere $\sum_1^5 m_ix_i=0$ are clearly equal to the expressions

$$-\Sigma\frac{m_if_i}{r_i}\Big/\Sigma\frac{m_i}{r_i},\quad -\Sigma\frac{m_ig_i}{r_i}\Big/\Sigma\frac{m_i}{r_i},\quad -\Sigma\frac{m_ih_i}{r_i}\Big/\Sigma\frac{m_i}{r_i};$$

where the point $(-f_i, -g_i, -h_i)$ is the centre of the fundamental sphere S_i.

Hence the coordinates of the centres of the set of spheres $m_i+\lambda m_i'$ are each of the form $\dfrac{P+\lambda Q}{R+\lambda S}$, i.e. the cross-ratio of any four of these points is equal to the cross-ratio of the corresponding values of λ.

Applying this to the spheres $\Sigma(a_i+\lambda)x_iy_i=0$, and taking λ to be successively $-\infty, -a_1 \ldots -a_5$, the corresponding centres are the point x and five points in which the normal to the cyclide at the point x meets the five fundamental quadrics $Q_1 \ldots Q_5$; it follows that the *cross-ratio of any four of the following six points on the normal at any point P of a cyclide is constant, viz. the point P and the centres of the bitangent spheres which lie on the normal at P.*

71. Confocal cyclides.

If the bitangent sphere is also a point-sphere z_i, its centre is a *focus* of the surface. Taking one system we have for instance

$$z_1=0,\quad z_i=(a_i-a_1)x_i,\qquad (i=2,\ldots 5),$$

with the condition

$$\sum_1^5 z_i^2=0.$$

The equations $z_1=0$, $\displaystyle\sum_2^5\frac{z_i^2}{a_i-a_1}=0$, give the focal curve.

If $\dfrac{1}{a_i+\mu}$ is substituted for a_i in the last equation, its form is not thereby altered, since a_i-a_1 is transformed into

$$\frac{1}{a_i+\mu}-\frac{1}{a_1+\mu}=\frac{a_1-a_i}{(a_i+\mu)(a_1+\mu)}.$$

Hence the equation becomes

$$\sum_2^5 \frac{a_i + \mu}{a_i - a_1} z_i^2 = 0,$$

which leads to the original equation

$$\sum_2^5 \frac{z_i^2}{a_i - a_1} = 0.$$

Hence the cyclides $\Sigma \frac{x_i^2}{a_i + \lambda} = 0$ are confocal with the original cyclide $\Sigma a_i x_i^2 = 0$. They form therefore a confocal system in which the original cyclide is included as corresponding to the value infinity for λ.

Through any point there pass three confocal cyclides, since the equation $\sum_1^5 \frac{x_i^2}{a_i + \lambda} = 0$, regarding the x_i as given, constitutes a cubic in λ (since $\sum_1^5 x_i^2 = 0$).

These cyclides cut each other orthogonally, for if λ_1, λ_2 refer to two cyclides through the point x, then since tangent spheres at this point to them respectively are

$$\Sigma \frac{x_i y_i}{a_i + \lambda_1} = 0, \quad \Sigma \frac{x_i y_i}{a_i + \lambda_2} = 0,$$

if these spheres are orthogonal we have

$$\Sigma \frac{x_i^2}{(a_i + \lambda_1)(a_i + \lambda_2)} = 0, \qquad \text{(Art. 61).}$$

But this is merely another form of the equation

$$\Sigma \frac{x_i^2}{a_i + \lambda_1} - \Sigma \frac{x_i^2}{a_i + \lambda_2} = 0.$$

The three cyclides through any real point are all real; for the variables x_i^2 may be all real, in which case the square of one of them, say x_5^2, is negative, so that if we suppose the quantities $a_1 \ldots a_4$ in order of magnitude, the cubic determining λ has a root in each of the intervals $-a_1 \ldots -a_2$, $-a_2 \ldots -a_3$, $-a_3 \ldots -a_4$. Again, if x_1 and x_2, and consequently a_1 and a_2, are conjugate imaginary, the cubic has a real root in each of the intervals $-a_3 \ldots -a_4$, $-a_4 \ldots -a_5$, and therefore possesses three real roots.

Corresponding points on confocal cyclides.

The equations

$$x_i = \frac{y_i}{\sqrt{a_i+\lambda}}, \qquad (i=1, \ldots 5),$$

establish a (1, 1) correspondence between the point x on the cyclide $\Sigma a_i x_i^2 = 0$ and the point y on the cyclide $\Sigma \frac{y_i^2}{a_i+\lambda} = 0$.

Denoting by $f(\lambda)$ the product $\prod_1^5 (\lambda + a_i)$, by resolving into partial fractions the expressions

$$\frac{\lambda^3}{f(\lambda)}, \quad \frac{\lambda^2}{f(\lambda)}, \quad \frac{\lambda}{f(\lambda)}, \quad \frac{1}{f(\lambda)},$$

it is seen that

$$\Sigma \frac{a_i^3}{f'(-a_i)} = \Sigma \frac{a_i^2}{f'(-a_i)} = \Sigma \frac{a_i}{f'(-a_i)} = \Sigma \frac{1}{f'(-a_i)} = 0.$$

Hence the equations $\Sigma a_i x_i^2 = 0$, $\Sigma x_i^2 = 0$, are identically satisfied by the substitutions

$$\rho x_i = \sqrt{\frac{(a_i+\lambda_1)(a_i+\lambda_2)}{f'(-a_i)}}, \qquad (i=1, \ldots 5).$$

These equations express the coordinates of any point x of the cyclide in terms of two parameters λ_1 and λ_2, so that if we take

$$x_i = \frac{y_i}{\sqrt{a_i+\lambda_3}},$$

it follows that

$$\rho y_i = \sqrt{\frac{(a_i+\lambda_1)(a_i+\lambda_2)(a_i+\lambda_3)}{f'(-a_i)}}, \quad (i=1, \ldots 5).$$

The quantities λ_1, etc. are seen to be the roots of the cubic in λ giving the three confocals through the point y.

The above expressions for the y_i in terms of $\lambda_1, \lambda_2, \lambda_3$ may be directly obtained by considering the cubic in the form

$$\frac{y_1^2}{\mu} + \sum_2^5 \frac{y_i^2}{\mu + a_i - a_1} = 0,$$

and hence it follows that

$$y_1^2 f'(-a_1) = \sum_1^5 a_i y_i^2 (a_1+\lambda_1)(a_1+\lambda_2)(a_1+\lambda_3);$$

four other equations of like form are obtained similarly.

The curves $\lambda_1 = \text{constant}$, $\lambda_2 = \text{constant}$ are the lines of curvature on the surface $\Sigma\, a_i x_i^2 = 0$, for from the equation

$$\rho x_i = \sqrt{\frac{(a_i + \lambda_1)(a_i + \lambda_2)}{f'(-a_i)}},$$

it follows that if $x_i + dx_i$ be the consecutive point on the curve $\lambda_2 = \text{constant}$, we have

$$\rho\, dx_i = -x_i d\rho + \frac{1}{2}\frac{\rho x_i d\lambda_1}{a_i + \lambda_1},$$

hence

$$(a_i + \lambda_1 - \tfrac{1}{2} d\lambda_1)(x_i + dx_i) = (a_i + \lambda_1)\, x_i \left(1 - \frac{d\rho}{\rho}\right),$$

neglecting quantities of the second order; therefore a tangent sphere at x_i, viz. $\Sigma\,(a_i + \lambda_1)\, x_i y_i = 0$, is also one of the tangent spheres at the consecutive point on the curve $\lambda_2 = \text{constant}$, and is therefore a *principal sphere* at the point x.

Thus the two confocals through any point of the surface $\Sigma\, a_i x_i^2 = 0$ intersect it in its lines of curvature; which is otherwise manifest from Dupin's theorem.

72. The sixteen lines of the surface.

It is known that every general quartic surface with a double conic contains sixteen lines (Art. 24). The existence of these lines on the cyclide is made evident by the equations

$$\rho x_i = \sqrt{\frac{(a_i + \lambda_1)(a_i + \lambda_2)}{f'(-a_i)}}, \quad (i = 1, \ldots 5) \ldots\ldots\ldots (1).$$

For if we suppose $\lambda_1 = \lambda_2$, we obtain for any point x of the curve $\lambda_1 = \lambda_2$ the equations

$$\rho x_i = A_i \lambda_1 + B_i, \qquad (i = 1, \ldots 5);$$

whence if ξ, η, ζ are the corresponding Cartesian coordinates of the point x,

$$\xi = \frac{\alpha_1 \lambda_1 + \alpha_2}{C\lambda_1 + D}, \quad \eta = \frac{\beta_1 \lambda_1 + \beta_2}{C\lambda_1 + D}, \quad \zeta = \frac{\gamma_1 \lambda_1 + \gamma_2}{C\lambda_1 + D};$$

hence the curve is a straight line.

By taking all combinations of signs in the ambiguities in equations (1) we obtain the sixteen lines. These lines are all imaginary, since as in the general case of the quartic surface with a nodal conic, a line on the surface must form part of a conic on the surface, and in the case of the cyclide all such conics are circles.

73. Centre of a cyclide.

The locus of mean distances of points of intersection of a series of parallel chords with an algebraic curve of degree n is called a *diameter*; when the terms of degree $n-1$ in the equation of the curve $f(x, y)=0$ are lacking, all diameters pass through the origin, the *centre* of the curve. The equation of a cyclide being written in the form

$$(x^2+y^2+z^2)^2+4U=0,$$

we proceed to consider the centres of its sections. Since the coordinate planes may have any directions, we may consider the section of the surface by the plane $z=h$; it is seen that the diameter corresponding to chords of the section parallel to the axis of x is the axis of y, and vice-versâ. Hence the line $x=y=0$ is the locus of centres of sections parallel to the plane $z=0$, so that the locus of centres of sections parallel to any plane is a line through the origin perpendicular to that plane. The origin is therefore termed the *centre of the surface.*

Sphero-conics on a cyclide.*

The sphere $S=2L$ (where $L\equiv \alpha x+\beta y+\gamma z+\delta$, $S\equiv x^2+y^2+z^2$) meets the cyclide $S^2+4U=0$ in a curve given as the intersection of the sphere and the quadric $U+L^2=0$; it is therefore a sphero-conic σ; the centre of the sphere is termed the centre of σ. Now denote by H the quadric $U+L^2+\frac{\lambda}{4}(S-2L)=0$; the intersection of H with the cyclide consists of σ together with another sphero-conic σ' which lies on the sphere $S+2L-\lambda=0$.

Hence *any quadric through σ meets the cyclide in another sphero-conic σ'.* The line joining the centres of σ and σ' is bisected at the centre of the surface, hence all quadrics through a given sphero-conic cut the cyclide in another sphero-conic whose centre is fixed.

If L is a constant k, the centres of σ and σ' coincide with the origin; if $4k=\lambda$, the spheres and therefore σ and σ' coincide, and H becomes the quadric V, where

$$V\equiv U+kS-k^2;$$

* The results of the present and following Articles were given by Humbert, *Sur les surfaces cyclides*, Journal de l'école polytechnique, LIV. (1884).

this quadric V *touches the cyclide along a sphero-conic lying on the sphere* $S-2k=0$. For these quadrics V it is easy to see that in general *two pass through any point, three touch any given line and four touch any given plane**.

If V correspond to any given value of k, and if in the equation of H previously given we suppose the quantity λ to be $4k$, it is seen that

$$H \equiv V + (L-k)^2;$$

hence H touches V along a conic.

The spheres Σ_1, Σ_2 which contain the curves σ, σ' of intersection of H and the cyclide are then

$$\Sigma_1 \equiv S - 2L, \quad \Sigma_2 \equiv S + 2L - 4k;$$

while the sphere Σ_3 passing through the curve of contact of V and the cyclide is $\Sigma_3 \equiv S - 2k$.

These three spheres have a circle in common in the plane $L-k=0$; hence, *every quadric H which cuts a cyclide in two sphero-conics σ, σ' touches one of the quadrics V along a conic; the spheres which respectively contain σ, σ' and the curve of contact of V and the cyclide have a circle in common whose plane is that of contact of V and H.*

Now take a generator of H through any point P of the conic $L-k=V=0$; this lies in the tangent plane of H at P and meets the curves σ, σ' in points A, B; C, D; also

$$PA \,.\, PB = \text{power of } P \text{ for } \Sigma_1, \quad PC \,.\, PD = \text{power of } P \text{ for } \Sigma_2,$$

and since P lies on the common radical plane of Σ_1, Σ_2, Σ_3 we have

$$PA \,.\, PB = PC \,.\, PD = \text{power of } P \text{ for } \Sigma_3 = x^2 + y^2 + z^2 - 2k;$$

where (x, y, z) are the coordinates of P.

By giving different values to the constants in L we obtain all quadrics H which touch V at P; hence the result holds for any tangent line to V at P, from which we deduce the result: *the point of contact of V with any of its tangent planes π is a centre of self-inversion for the section of the cyclide by π*; and since four quadrics V can be drawn to touch any given plane we thus obtain the four centres of self-inversion of the section of the cyclide by any plane.

* The quadrics V correspond to the quadrics $\psi + 2\lambda\phi + \lambda^2 w^2 = 0$ for the general quartic surface $\phi^2 = w^2\psi$.

74. Conjugate points.

Two such points A, B on a tangent to V are called *conjugate* for the system V. Any two points of the cyclide are conjugate for one quadric V and for one only; for since three quadrics V touch any given line which meets the cyclide in the points A, B, C, D, these points can be arranged in two pairs in three ways, each arrangement corresponding to one quadric V.

If from any point A of the cyclide tangents be drawn to a given quadric V the conjugates of A lie upon the cone and also upon the cyclide; since the cone has as its equation $VV' - P^2 = 0$ where P is the polar plane of A (or x', y', z'), it follows that this cone meets the cyclide in two sphero-conics σ, σ' but only one of them is formed by the conjugates of A. For since $V \equiv U + kS - k^2$ we obtain the intersection of this cone with the surface $S^2 + 4U = 0$ by writing in the equation of the cone $4V = -(S-2k)^2$, giving as the two spheres through the curves σ, σ'

$$(S-2k)(S'-2k) = \pm 4P;$$

one of these spheres passes through (x', y', z'), since

$$P' = V' = -\frac{(S'-2k)^2}{4}.$$

Let Σ_1 be the sphere which passes through A, then if Q is any point (x, y, z) on the conic along which the cone touches V, and therefore lying on the plane $P = 0$, the line AQ meets Σ_1 in a second point B such that

$$QA \cdot QB = \text{power of } Q \text{ for } \Sigma_1 = x^2 + y^2 + z^2 - 2k.$$

Hence B is conjugate to A; and σ is composed of the conjugates of A.

The direction cosines of the normals to the cyclide and the sphere Σ_1, at A, are easily seen to be the same. Hence *the locus of the conjugates of a point A for a given quadric V lies on a sphere touching the surface at A.*

75. Cartesian equation of the system of confocal cyclides.

The equation $S^2 + \dfrac{4F(\lambda)Q}{(A_1+\lambda)(A_2+\lambda)(A_3+\lambda)} = 0$, where

$$S \equiv x^2 + y^2 + z^2 + \frac{2B_1x}{A_1+\lambda} + \frac{2B_2y}{A_2+\lambda} + \frac{2B_3z}{A_3+\lambda} + 2\lambda,$$

$$Q \equiv \frac{x^2}{A_1+\lambda} + \frac{y^2}{A_2+\lambda} + \frac{z^2}{A_3+\lambda} + 1, *$$

* See Art. 54.

represents in Cartesian coordinates the system confocal to the given cyclide. For it is, when reduced, of degree eight in λ, and since

$$F(-A_1) = -B_1^2(A_2 - A_1)(A_3 - A_1),$$

it is seen that $\lambda + A_1$ is a factor of the reduced equation; similarly for $\lambda + A_2$, $\lambda + A_3$. Moreover the coefficients of λ^8 and λ^7 vanish, hence we have a resulting cubic λ of the form

$$\lambda^3\Sigma_1 + \lambda^2\Sigma_2 + \lambda\Sigma_3 + \Sigma_4 = 0,$$

where the Σ_i are cyclides. Since for the roots $\lambda_1 \ldots \lambda_4$ of $F(\lambda) = 0$ the surface reduces respectively to $S_1^2 \ldots S_4^2$, it follows that the cubic in λ can be expressed in the form

$$\sum_1^4 \alpha_i S_i^2 = 0,$$

where the α_i are cubic in λ; and therefore in the form

$$\frac{\kappa_1\left(\dfrac{S_1}{r_1}\right)^2}{\lambda - \lambda_1} + \ldots + \frac{\kappa_4\left(\dfrac{S_4}{r_4}\right)^2}{\lambda - \lambda_4} = 0.$$

But since S_5 is included, for $\lambda = \lambda_5$, it follows that

$$\kappa_1 : \kappa_2 : \kappa_3 : \kappa_4 = \lambda_5 - \lambda_1 : \lambda_5 - \lambda_2 : \lambda_5 - \lambda_3 : \lambda_5 - \lambda_4,$$

whence we finally obtain $\displaystyle\sum_1^5 \frac{\left(\dfrac{S_i}{r_i}\right)^2}{\lambda - \lambda_i} = 0.$

The following result is given by Humbert*; *when the sphere* $S + 2L = 0$ *is a point-sphere, and the quadric* $U + L^2 = 0$ *is a cone, the locus of the centre of this point-sphere is a cyclide confocal with* $S^2 + 4U = 0$.

Let

$$S \equiv x^2 + y^2 + z^2 + d,$$
$$U \equiv a_1x^2 + a_2y^2 + a_3z^2 + 2px + 2qy + 2rz + c,$$
$$L \equiv \alpha x + \beta y + \gamma z + \delta.$$

Then $S + 2L = 0$ is a point-sphere if

$$\alpha^2 + \beta^2 + \gamma^2 - d - 2\delta = 0;$$

the quadric $U + L^2 = 0$ is a cone if

$$\alpha^2\Delta_{11} + 2\alpha\beta\Delta_{12} + \ldots + \Delta = 0,$$

where Δ is the discriminant of U, and Δ_{11}, etc. its first minors.

* *Sur les surfaces cyclides*, Journal de l'école polytechnique, LIV. (1884).

Now denoting by (x, y, z) the centre of the point-sphere so that $x = -\alpha$, etc. the last equation becomes

$$\left(\delta + \frac{px}{a_1} + \frac{qy}{a_2} + \frac{rz}{a_3}\right)^2 + \frac{\Delta}{a_1 a_2 a_3}\left(\frac{x^2}{a_1} + \frac{y^2}{a_2} + \frac{z^2}{a_3} + 1\right) = 0.$$

On inserting the value of δ we obtain as the required locus

$$\left(x^2 + y^2 + z^2 + \frac{2px}{a_1} + \frac{2qy}{a_2} + \frac{2rz}{a_3} - d\right)^2 + \frac{4\Delta}{a_1 a_2 a_3}\left(\frac{x^2}{a_1} + \frac{y^2}{a_2} + \frac{z^2}{a_3} + 1\right) = 0.$$

On writing $a_i = A_i + \lambda$, $p = B_1$, $q = B_2$, $r = B_3$, $d = -2\lambda$, $c = C - \lambda^2$, the cyclide $S^2 + 4U = 0$ takes the form given (Art. 54) and the locus is seen to be a confocal cyclide.

76. Common tangent planes of the cyclide and a tangent quadric.

If we take any plane touching the cyclide at a point O and an inscribed quadric V at a point P, the line PO touches a line of curvature of the cyclide at O. For if the plane be taken as the plane $z = 0$ and the line PO as the axis of x, the equation of the surface assumes the form

$$(x^2 + y^2 + z^2 + 2ax + 2by + 2cz + k)^2$$
$$+ 4\left(Ax^2 + By^2 + Cz^2 + 2Dyz + 2Ezx + 2Fxy\right.$$
$$\left. - kax - kby + 2rz - \frac{k^2}{4}\right) = 0,$$

with the conditions $F + ab = 0$, $A + a^2 = 0$; the second member of the left side representing V.

But in the equation of the indicatrix of the surface at O the coefficient of the term involving xy is $F + ab$, hence the line PO is a tangent to a line of curvature at O. The tangent to the other line of curvature is OQ, where Q is the point of contact of the other inscribed quadric V' which can be drawn to touch the plane (the two other inscribed quadrics which touch the plane coincide here with that which touches the cyclide at O taken *doubly*).

Thus the locus of the points of contact with a cyclide of a plane which touches the cyclide and a fixed quadric V, is a line of curvature of the cyclide, the intersection with it of a confocal cyclide.

CHAPTER VI

SURFACES WITH A DOUBLE LINE: PLÜCKER'S SURFACE

77. The equation of a quartic surface with a double line may be written in the form

$$x_1^2U + 2x_1x_2V + x_2^2W = 0,$$

where $U=0$ is a quadric and $V=0$, $W=0$ are cones whose vertex is the point A_1. Since twenty-two constants enter linearly into this equation, and since four conditions determine a line, the surface depends upon twenty-five constants.

The section of the surface by a plane $x_1 = \lambda x_2$ through the double line consists of the double line together with a conic; the cone of vertex A_1 through this conic has as its equation

$$\lambda^2U' + 2\lambda V + W = 0,$$

where U' is the result of substituting λx_2 for x_1 in U. The coefficients of x_2^2, x_3^2, x_4^2, x_2x_3, x_2x_4, x_3x_4 in the last equation are functions of λ of degrees 4, 2, 2, 3, 3 and 2 respectively; hence it is a pair of planes for eight values of λ, eight of the sections through the double line consisting of a pair of lines. The surface thus contains sixteen lines; it contains no line which does not intersect the double line, unless U, V, W have a common generator.

78. In addition to the conics in planes through the double line the surface contains certain other conics. The origin of these conics is seen by application of the following theorem: *if seven lines $p_1p_2 \ldots p_7$ are all intersected by an eighth line p, there is one conic which intersects the eight lines.*

This result may be proved as follows: consider five lines $p_1 \ldots p_5$ all intersected by p which we may take as the edge A_3A_4, any arbitrary line p' being A_1A_2, *then there are three planes through p' which meet the lines $p_1 \ldots p_5$, A_3A_4 in points of a conic*;

for the plane $x_3 - \lambda x_4 = 0$ meets any one of the lines $p_1 \dots p_5$ in a point whose coordinates are

$$A\alpha,\ B\alpha,\ \lambda,\ 1,$$

where $\alpha = a\lambda + b$ and a, b, A, B are constants connected with the line; project these five points from A_4 upon the plane α_4, giving points

$$A_i\alpha_i,\ B_i\alpha_i,\ \lambda, \qquad (i = 1, 2, 3, 4, 5);$$

the condition that these five points and A_3 should lie upon a conic is then

$$\begin{vmatrix} A_1^2\alpha_1^2 & B_1^2\alpha_1^2 & \lambda^2 & A_1\lambda\alpha_1 & B_1\lambda\alpha_1 & A_1B_1\alpha_1^2 \\ \vdots & \vdots & \vdots & \vdots & \vdots & \vdots \\ 0 & 0 & 1 & 0 & 0 & 0 \end{vmatrix} = 0.$$

Omitting the irrelevant factor λ^2 we obtain an equation of degree eight in λ. But five of these values of λ relate to the five planes through p' and the points (p_i, A_3A_4); hence we have finally three planes through p' meeting $p_1 \dots p_5$, A_3A_4 in points of a conic.

Hence the planes meeting p, $p_1 \dots p_5$ in points of a conic envelop a surface of the third class.

This surface contains each of the lines $p_1 \dots p_5$; for the plane through p_1 and the second transversal of p_1, p_2, p_3, p_4, meets the lines $p, p_1 \dots p_5$ in six points lying on two lines.

Similarly for p_1 and the lines p_3, p_4, p_5, etc.; hence four tangent planes of the surface can be drawn through p_1, i.e. p_1 lies on the surface; similarly for $p_2 \dots p_5$. For the same reason p lies on the surface. Now consider the three surfaces thus formed with p, p_1, p_2, p_3, p_4 and p_5, p_6, p_7 respectively; applying the known results* for the intersections of three cubic surfaces which have four lines in common, it is seen that there is *one* tangent plane common to the three surfaces.

Construct, therefore, the conic which meets the double line and also one line of each pair out of seven pairs of lines; this conic meets the surface in nine points and therefore lies upon it. The plane of this conic meets the surface in another conic, the two conics have one intersection upon the double line; each conic meets *one* of the two lines forming the eighth pair of lines.

By taking all possible selections of seven lines in accordance with the foregoing method, we obtain $2^7 = 128$ conics lying in 64 planes; each plane being a tritangent plane of the surface.

* Salmon, *Geom. of three dimensions*, 5th Ed., Vol. I. p. 371.

79. Mapping of the surface on a plane.

Any one of the foregoing conics affords a means of representing the points of the surface upon a plane*. For if $C=0$ is such a tritangent plane, the equation of the surface may be written

$$(x_1 A - x_2 B)(x_1 A' - x_2 B') - C\{Px_1^2 - 2Nx_1 x_2 + Mx_2^2\} = 0,$$

where $A, B, \ldots M$ are linear functions of the coordinates.

We may therefore express any point of the surface in terms of two parameters, viz. the ratios of ξ_1, ξ_2, ξ_3, as follows:

$$(1)\quad \xi_1 x_1 + \xi_2 x_2 = 0,$$

$$(2)\quad \xi_1 B + \xi_2 A + \xi_3 C = 0,$$

$$(3)\quad \xi_3(B'\xi_1 + A'\xi_2) + M\xi_1^2 + 2N\xi_1\xi_2 + P\xi_2^2 = 0;$$

giving a (1, 1) correspondence between any point x of the surface and a point ξ of any assumed plane.

For any assigned point ξ, the first two equations give a line which intersects the double line and also the conic

$$C=0, \quad x_1 A - x_2 B = 0,$$

its fourth point of intersection with the surface being the point x which corresponds to ξ. Conversely each point x of the surface determines such a line and hence one point ξ. For any point x, however, of one of the eight lines of the surface which intersect this conic, the same point ξ is determined; we have therefore eight *principal points* ξ of the correspondence which we denote by $B_1 \ldots B_8$, each of them corresponds to all the points of one of the eight lines.

If in the preceding equations connecting x and ξ we have $C=0$, then either $x_1 A - x_2 B = 0$, or $\xi_1 = 0$, $\xi_2 = 0$; hence to the points of the conic in the plane $C=0$ which does not meet the line determined by (1) and (2), there corresponds the single point $\xi_1 = 0$, $\xi_2 = 0$, which we denote by A.

To any plane section of the surface there corresponds in the field of ξ a quartic curve: since this section meets each of the above eight lines, and also twice meets the conic just referred to, it follows that this quartic curve passes through the eight principal points and also passes twice through the point A. Hence we have a system of quartics having a common node and eight common points†.

* Clebsch, *Ueber die Abbildung algeb. Flächen*, Math. Ann. I.

† The condition of possessing a node at a given point and eight fixed points is equivalent to eleven conditions, leaving a linear system triply infinite of quartic curves.

To the conic $C=0$, $\xi_1 x_1+\xi_2 x_2=0$, $\xi_1 B+\xi_2 A=0$ there corresponds a quartic curve obtained by substituting for the coordinates x their values in (3), in terms of $\xi_1:\xi_2$; hence this quartic possesses a triple point at A.

The pencil of lines through the point A corresponds to the conics in sections through the double line; the cubic curve through the nine principal points corresponds to the double line. This cubic is obtained by writing $x_1=x_2=0$ in (2) and (3), and hence is given as the intersection of a pencil of lines $x_3K+x_4K'=0$, and a pencil of conics $x_3U+x_4U'=0$; to any given point x_3/x_4 of the double line correspond two points P, P' of this cubic collinear with the point $K=K'=0$, or O.

Writing $x_1=x_2=0$ in (2) and (3) and eliminating ξ_3 we obtain

$$(\xi_1B_0+\xi_2A_0)(\xi_1B_0'+\xi_2A_0')-C_0(P_0\xi_2^2+2N_0\xi_1\xi_2+M_0\xi_1^2)=0,$$

in which B_0 is the result of writing $x_1=x_2=0$ in B, etc.

This gives the pair of lines joining P, P' to the point $\xi_1=\xi_2=0$; the corresponding pair of planes through the double line is

$$(x_1A_0-x_2B_0)(x_1A_0'-x_2B_0')-C_0(P_0x_1^2-2N_0x_1x_2+M_0x_2^2)=0,$$

which is the pair of tangent planes at the stated point x_3/x_4 of the double line. Hence the conics in these tangent planes are so related that their corresponding lines meet the cubic, the image of the double curve, in points P, P' collinear with the point O.

From O four tangents can be drawn to this cubic, hence at four points of the double line the tangent planes coincide, giving four *pinch-points*.

80. We add a table containing the preceding results:

On the surface	In the field of ξ
Eight lines of the surface which meet the conic $C=x_1A-x_2B=0$, or c_2^2	Eight principal points lying on a quartic having a triple point at $\xi_1=\xi_2=0$
The second conic in the plane $C=0$, or c_1^2	A principal point A ($\xi_1=\xi_2=0$)
The conics in the planes through the double line	Lines of the pencil whose centre is the point A
The double line	A cubic passing through the nine principal points
A point Q on the double line	Two points P, P' in this cubic collinear with a fixed point O on the cubic

On the surface	In the field of ξ
The conics in the pair of tangent planes to the surface at Q	Two lines AP, AP'
Four pinch-points on the double line	The four points of contact of the tangents from O to the cubic

The nine principal points cannot be the complete intersection of two cubic curves $u=0$, $v=0$; for if so the image of every plane section would be of the form $Lu+Mv=0$, where L, M, u, v concur at A. Hence the equations connecting x and ξ would be of the form

$$\rho x_1 = Pu, \quad \rho x_2 = Pv, \quad \rho x_3 = Qu, \quad \rho x_4 = Qv,$$

leading to the quadric surface $x_1 x_4 = x_2 x_3$.

The curve on the surface which corresponds to any line in the plane of ξ is a twisted quartic of the second species. For to any line $a_\xi = 0$ of the plane there will correspond a curve lying on the quadric

$$\begin{vmatrix} a_1 & a_2 & a_3 \\ x_1 & x_2 & 0 \\ B & A & C \end{vmatrix} = 0;$$

this quadric meets the quartic surface in the double line and also in the conic $c_2{}^2$; hence it also meets it in a twisted quartic. Since the line $a_\xi = 0$ meets the cubic corresponding to the double line three times, the double line meets the quartic curve three times; this quartic is therefore of the second species.

To any line of the pencil whose centre is the point O of the cubic curve corresponds a quartic which passes through the point of the double line corresponding to O and which has a double point in the single point corresponding to the points P, P'.

81. Curves on the surface.

By aid of this representation of the surface on a plane the various algebraic curves on the surface can be readily determined.

If M and m are the orders of a curve c_x on the surface and the corresponding curve c_ξ in the plane, c_x is met by any plane section a_x in M points, and c_ξ is met by a_ξ in $4m$ points; while if c_x meets $c_1{}^2$ β times* and the eight lines $\alpha_1 \ldots \alpha_8$ times

* Excluding an intersection at the point where $c_1{}^2$ meets the double line.

respectively, these points, though not intersections of c_x and a_x, give rise to intersections of c_ξ and a_ξ, hence

$$M = 4m - 2\beta - \Sigma\alpha.$$

By aid of this equation we can obtain the various species of curves on the surface*.

The *sixteen lines* are represented by the eight principal points B_i and the lines joining the point A to these eight points.

The *conics* of the surface are obtained by taking $\beta = 0, 1, 2, 3$ successively:

$\beta = 0$ requires that $m = 1$, $\Sigma\alpha = 2$; this gives the joins of the eight principal points B_i, which are twenty-eight in number.

$\beta = 1$ requires that $m = 1$ or $m = 2$; in the first case $\Sigma\alpha = 0$ and we have the pencil of lines through the point A; in the second case $\Sigma\alpha = 4$ and we have conics through A and four principal points; there are seventy such conics.

$\beta = 2$ requires that $m = 3$; this gives $\Sigma\alpha = 6$, and we have thus cubics having a node at A and passing through six principal points; there are twenty-eight such cubics.

$\beta = 3$ requires that $m = 4$ and hence $\Sigma\alpha = 8$; this gives a quartic having a triple point at A and passing through the eight principal points. This quartic corresponds to c_2^2.

The case $\beta = 4$ cannot arise. We have thus, counting the point A, obtained the images of all the conics of the surface, including those in sections through the double line; apart from the latter there are

$$1 + 28 + 70 + 28 + 1 = 128$$

conics on the surface; i.e. there are no conics other than those already obtained.

* Limits within which m must lie are derived from the fact that $m \geqslant \beta + 1$ and from the equation

$$p = \frac{(m-1)(m-2)}{2} - \frac{\beta(\beta-1)}{2} - \Sigma\frac{\alpha(\alpha-1)}{2},$$

where p is the deficiency of the plane curve, if we suppose the curve on the surface not to possess multiple points; and also from the inequality

$$1 \leqslant \frac{(m+1)(m+2)}{2} - \frac{\beta(\beta+1)}{2} - \Sigma\frac{\alpha(\alpha+1)}{2}.$$

It follows that

$$M + p - 1 \geqslant 4m - 3m - \beta,$$

hence

$$m \leqslant M + p - 1 + \beta.$$

Cubics on the surface.

The case $M=3$ gives rise to curves in the plane of which particulars are stated in the following table. We notice that since the coordinates of a point on a twisted cubic are expressible as cubic functions of a parameter, the corresponding curve in the plane is unicursal, so that for it $p=0$.

	β	$\Sigma\alpha$	m	
I	0	1	1	Eight pencils of lines through $B_1 \dots B_8$
II	0	5	2	56 conics through five of the points B_i
III	1	3	2	56 sets of ∞^1 conics through A and three points B_i
IV	1	7	3	168 cubics through A and five points B_i and having a sixth point B_i as node
V	2	5	3	56 sets of ∞^1 cubics with a node at A and passing through five points B_i
VI	2	9	4	168 quartics having nodes at A and at two points B_i and passing through five other points B_i
VII	3	7	4	Eight sets of ∞^1 quartics having a triple point at A and passing through seven points B_i
VIII	3	11	5	56 quintics having a triple point at A, passing through five points B_i and having nodes at three points B_i

Any conic of class III intersects the cubic corresponding to the double line in two points* apart from the principal points, hence the corresponding cubic on the surface meets the double line twice. The same applies to the cubics corresponding to class V. And any conic of class III meets any cubic of class V in four points apart from A, if they together pass through the eight points B_i. Hence if c^3, c'^3 are any two such cubics on the surface represented in classes III and V respectively, we can pass *one* quadric through the double line, their points of intersection, and one point on each respectively†; this quadric will contain each cubic. It follows therefore that *any two cubics thus represented by classes III and V respectively lie on the same quadric; there are ∞^2 quadrics which meet the surface in the double line and two twisted cubics.*

Similarly any two curves of classes I and VII which together pass through the eight points B_i will intersect in four points and meet the cubic corresponding to the double line in two points. Hence they also lie on one of ∞^2 quadrics. The 448 simple cubics in classes II, IV, VI and VIII arise as the intersection with the surface of the 448 quadrics which pass through three non-intersecting lines of the surface and the double line.

* These two points are not collinear with O as is seen by taking the pair of lines AB_1, B_2B_3 and the corresponding line and conic on the surface.

† Any quadric through the double line contains six available constants.

Quartic curves of each species exist on the surface: their images in the plane will have deficiency either zero or unity. The following table gives the varieties of such images:

$\beta=0$	$\beta=1$	$\beta=2$
$m=1,\ \Sigma a=0$	$m=2,\ \Sigma a=2$	$m=3,\ \Sigma a=4$
$m=2,\ \Sigma a=4$	$m=3,\ \Sigma a=6$	$m=4,\ \Sigma a=8$
$m=3,\ \Sigma a=8$	$m=4,\ \Sigma a=10$	$m=5,\ \Sigma a=12$
		$m=6,\ \Sigma a=16$

The quartics which arise for $\beta=4$, $\beta=3$, are similar to those for which $\beta=0$, $\beta=1$ respectively. The quartics lie in pairs on quadrics; if β, m, $\Sigma\alpha$ and β', m', $\Sigma\alpha'$ correspond to such a pair of quartic curves lying on the same quadric, we have

$$\beta+\beta'=4, \quad m+m'=8, \quad \Sigma\alpha+\Sigma\alpha'=16.$$

By applying the method to quintic curves it is seen that the image of every quintic curve on the surface meets the cubic which is the image of the double line in at least one point apart from the principal points: hence every quintic curve on the surface meets the double line at least once. Since a triply infinite number of cubic surfaces can be drawn through any quintic curve, it follows that at least one cubic surface passes through the double line and any quintic on the surface, which it therefore meets also in another quintic curve. Hence the quintics lie in pairs on cubic surfaces through the double line.

Since one cubic surface passes through any twisted sextic curve the sextics on the surface lie in pairs on cubic surfaces.

82. Nodes on the surface.

Many of the preceding results, and also the modifications which arise when the surface contains nodes apart from the double line, may be investigated by means of Rohn's method, Chap. I. Taking the equation of the surface as being

$$x_1^2 U + 2x_1 V + W = 0,$$

where the double line is the join of $A_1 A_2$; U, V, W are quadratic in x_3, x_4; and x_2 appears in the first degree in V and in the second degree in W. The tangent cone to the surface from A_1 is then $V^2 - UW = 0$; its section by $x_1 = 0$ is a sextic curve having a quadruple point at A_2. The class of this sextic is $30 - 12 = 18$, hence ten tangents can be drawn to it from A_2. The two tangents from A_1 to the conic in any plane section through the double line touch it in points P, P' which are projected from A_1 into points Q, Q' collinear with A_2; such a pair of points Q, Q' coincide, as just seen, ten times; such a coincidence arises from the pair of conics which pass though A_1, these two tangents are the lines $U = 0$, they lie in the two tangent planes to the surface at A_1; and also when a conic becomes a pair of lines; hence there are eight pairs of lines meeting the double line.

If the surface has a node external to the double line, a node arises on this sextic curve, hence its class is reduced by two, and the number of pairs of lines is reduced by unity.

If there are five nodes the sextic necessarily breaks up into a quintic having a triple point at A_2 and three nodes, together with a line through A_2; for it cannot break up into a quartic with a triple point and a conic, or into two cubics with a common double point, since in both cases no tangent can be drawn to the compound curve to touch it at a point outside A_2, while there should be two, corresponding to the two conics through A_1.

Hence two nodes must lie in a plane through the double line, and their join meets the double line. The tangent cone from A_1 breaks up into a quintic cone together with the plane through the double line and the line p joining the nodes. This plane touches the surface along the line p: there is no part of the surface in this plane except the double line and p; for any line in the plane meets the surface twice where it meets p and twice where it meets the double line and therefore at no other point; the line p is therefore *torsal*.

If there are six nodes the sextic becomes a trinodal quartic having one of its nodes at A_2 and two lines through A_2; here we have two torsal lines. If there are seven nodes the sextic becomes a nodal cubic through A_2 and three lines through A_2; finally, if there are eight nodes, we have a conic and four lines through A_2.

It should be noticed that if there are four coplanar nodes their plane is a trope of the surface. For the conic through the four nodes and the point in which the plane meets the double line meets the surface in ten points, and hence lies wholly on the surface. The two conics of the section here become coincident, and the plane is a trope.

Consider next the case of seven nodes; six of them, as has been seen, lie in pairs on three lines meeting the double line, let them be B_1, B_1' lying on b_1, B_2, B_2' lying on b_2, B_3, B_3' lying on b_3, and a seventh node B_4. Take as the point A_1 that in which the plane $B_2B_3B_4$ meets the double line; the sextic curve consisting as above, of a cubic through A_2 with one node, the projection of B_4, together with three lines through A_2, will for this position of A_1 have its node collinear with the projections of B_2 and B_3; hence it must consist in part of the line joining these two points, and upon this line the projection of another node, B_1', must lie. The residual part of the cubic is a conic. Hence it follows that the nodes $B_1'B_2B_3B_4$ are coplanar.

Taking the other combinations

$$B_1B_2B_3'B_4, \quad B_1B_2'B_3B_4, \quad B_1'B_2'B_3'B_4,$$

we obtain in all four planes, each containing four nodes. The surface has therefore four tropes.

If there is an eighth node B_4', then the sextic curve, consisting in the previous case of three lines through A_2 and a nodal cubic, here receives an additional double point, the projection of B_4', hence the sextic consists of four lines through A_2 and a conic. There are therefore four torsal lines, viz. B_1, B_1' on b_1, B_2, B_2' on b_2, B_3, B_3' on b_3, and B_4, B_4' on b_4.

By combining B_4' with the nodes $B_1 \ldots B_3'$ as before, we obtain four more tropes, viz. the planes

$$B_1B_2B_3B_4', \quad B_1'B_2B_3'B_4', \quad B_1'B_2'B_3B_4', \quad B_1B_2'B_3'B_4'.$$

It follows that the points B_i and the points B_i' form two tetrahedra, each of which is inscribed in the other.

If we take as the point A_1 that in which $B_2B_3B_4'$ meets the double line, the sextic curve consisting of four lines through A_2 and a conic will have three collinear nodes on the latter, viz., the projections of B_2, B_3, and B_4', hence this conic will consist of two lines, each containing four nodes, viz. those lying in the tropes

$$(B_1B_2B_3B_4') \text{ and } (B_1'B_2'B_3'B_4).$$

The line of intersection of these tropes thus meets the double line in A_1; and since the tangents from A_2 to this (degenerate) conic coincide, it follows that the tangent planes at A_1 will coincide and A_1 is a pinch-point. Hence we have four pairs of tropes whose four lines of intersection meet the double line at pinch-points.

83. Plücker's surface.

The surface with a double line and eight nodes is known as *Plücker's surface.* One form of its equation may be obtained as follows: through two nodes on different torsal lines there pass two tropes, and there is one quadric S which contains the conics in which these tropes meet the surface and also the double line. If the double line is A_1A_2 and the tropes are taken as the planes x_1 and x_2, the equation of the surface is necessarily of the form

$$S^2 + x_1x_2\gamma\delta = 0,$$

where $\gamma = 0$, $\delta = 0$ are two planes through the double line also meeting the surface in torsal lines.

Taking the two nodes as the vertices A_3 and A_4 of reference we obtain the equation of the surface in the form

$$(x_1\alpha + x_2\beta + nx_3x_4)^2 + x_1x_2\gamma\delta = 0,$$

where α, β, γ, δ are planes through the double line.

There are two further conditions to be satisfied, viz. that each of the planes x_3 and x_4 meets the surface, apart from the double line, in a torsal line; hence if

$$\alpha = a_3x_3 + a_4x_4, \quad \beta = b_3x_3 + b_4x_4, \quad \gamma = c_3x_3 + c_4x_4, \quad \delta = d_3x_3 + d_4x_4,$$

then writing successively $x_4 = 0$ and $x_3 = 0$, we obtain as the required conditions

$$4a_3b_3 = -c_3d_3, \quad 4a_4b_4 = -c_4d_4,$$

giving two torsal lines as

$$x_4 = x_1a_3 - x_2b_3 = 0, \quad x_3 = x_1a_4 - x_2b_4 = 0;$$

the other two being

$$\gamma = 0 = S, \quad \delta = 0 = S.$$

The above form of equation of the surface shows that the plane $x_1 = 0$ touches the conics lying in sections through the double line.

Hence each trope touches each of these conics.

By aid of this form of the equation of the surface we can obtain the form assumed in the case of Plücker's surface by the previous equations connecting a point x of the surface with a point ξ of a plane; for the equation of the surface is seen to be identically satisfied if we write

$$\rho x_1 = n\xi_1\xi_2 CD,$$
$$\rho x_2 = -n\xi_1\xi_2\xi_3^2,$$
$$\rho x_3 = \xi_1\{\xi_3^2 B + \xi_3 CD - ACD\},$$
$$\rho x_4 = \xi_2\{\xi_3^2 B + \xi_3 CD - ACD\},$$

where A, B, C, D are the results of substituting ξ_1 and ξ_2 for x_3 and x_4 respectively in α, β, γ and δ.

The cubic corresponding to the double line is

$$\xi_3^2 B + \xi_3 CD - ACD = 0;$$

it touches the lines $C = 0$, $D = 0$, and by aid of the preceding conditions it follows that it also touches the lines

$$\xi_1 = 0, \quad \xi_2 = 0.$$

The quartic representing any plane section has a node at the point $\xi_1 = \xi_2 = 0$, and touches the lines $\xi_1 = 0$, $\xi_2 = 0$ in given points. It also touches the lines $C = 0$, $D = 0$ where they respectively meet $\xi_3 = 0$. If $E_1E_2E_3$ is the triangle of reference in the field of ξ and the principal points on E_3E_2, E_2E_1 are Q_1 and Q_2, those on E_1E_2 are Q_3 and Q_3', the correspondence between points x and ξ is of the $(1, 1)$ character with the following exceptions: any point on E_3E_2 corresponds to the node A_4 with the exception of Q_1, which corresponds to a torsal line, and the point E_3 which corresponds to *any* point of the conic in the trope x_1; similarly for E_1E_3 and the point Q_2; while to any point of $C = 0$, $D = 0$ correspond the other two nodes in x_1 respectively, but to Q_3 and Q_3' the other two torsal lines respectively correspond.

The equation

$$M = 4m - 2\beta - \Sigma\alpha$$

as before connects the order of a curve on the surface and the corresponding curve in the plane, where β is the number of intersections of the former curve with the conic in the plane x_1 *exclusive* of intersections at nodes in that plane.

It is easily seen from this equation that no line can exist on the surface except the four torsal lines, which are represented by the four principal points Q, and the double line. To obtain the conics of the surface we may take $\beta = 0$ giving $m = 1$, $\Sigma\alpha = 2$; so that six conics are represented by the lines joining the principal points Q_1, Q_2, Q_3, Q_3'; or we may take $\beta = 1$, giving either

$m=1$, $\Sigma\alpha=0$, so that the conics in planes through the double line are represented by the pencil of lines through E_3; or $m=2$, $\Sigma\alpha=4$, which gives the conic through the five principal points. If we add the conic represented by E_3 we obtain all the conics on the surface.

The existence on the general surface with a double line, of conics whose planes do not contain the double line, and the fact that the surface is rational, have been shown very simply by Baker* as follows: consider the quadric cones

$$y=\frac{u}{w},\quad t=\frac{v}{w}\quad \text{......................}(1),$$

where $u=0$, $v=0$ are pairs of planes through the double line $x=z=0$; and $w=0$ is a plane through it; these cones intersect in a conic which meets the double line at one point only.

The equation of the surface being

$$x^2A+xzB+z^2C=0,$$

if in A, B, C we substitute for y and t from (1), we obtain a sextic in x/z; and the seven arbitrary constants in u, v and w may be so chosen as to make this sextic vanish identically. Hence the surface contains at least one conic which meets the double line once only.

The substitution $t=\frac{v}{w}+\tau\left(y-\frac{u}{w}\right)$ enables us to express y rationally in terms of τ and $\frac{x}{z}$; and hence also t; i.e. we can express the coordinates of any point on the surface as rational functions of two variables.

* *Some recent advances in the theory of algebraic surfaces*, Proc. Lond. Math. Soc., Series 2, Vol. XII. p. 36.

CHAPTER VII

QUARTIC SURFACES WITH AN INFINITE NUMBER OF CONICS: STEINER'S SURFACE: THE QUARTIC MONOID

84. The property of containing an infinite number of conics has been seen to be possessed by all quartic surfaces with a double conic or a double line; in the present chapter we consider all surfaces which have this property.

The determination of the quartic surfaces which contain an infinite number of conics was made by Kummer*.

The following is a brief account of his investigation. If the plane section of a quartic surface has four double points it will consist either of two conics or of a line together with a nodal cubic, in the latter case three double points are collinear. If the section consists of two conics, each of their points of intersection is either a double point of the surface or a point of contact of the surface and the plane section.

Consider first the case in which no point is a point of contact. Let *two* of the double points be fixed†; the surface must then possess a double conic, the equation of such a surface is

$$\Phi^2 = 4p^2\Psi,$$

where $\Phi = 0$, $\Psi = 0$ are quadrics. If the surface contain also two double points whose join does not lie on the surface, Ψ must break up into two linear factors (Art. 38), and the surface is then

$$\Phi^2 = 4p^2qr,$$

whose sections by planes through the line (q, r) are pairs of conics.

If the surface has another pair of nodes, its equation will be (Art. 38)

$$(p^2 + qr - st)^2 = 4p^2qr,$$

or

$$(p^2 - qr + st)^2 = 4p^2st.$$

* *Ueber die Flächen vierten Grades auf welchen Schaaren von Kegelschnitte liegen*, Crelle, LXIV.

† The cases in which none or only one of the double points are fixed lead only to a quartic surface consisting of two quadrics, or a cone.

In this case we have two sets of conics lying in planes whose axes are (q, r) and (s, t).

If three of the double points are fixed they necessarily lie on a double line; the sections of a surface having a double line by planes through the double line form a set of conics.

If certain of the double points coincide we are led to special cases of the surface $\Phi^2 = 4p^2qr$, except in one case, viz. that in which the surface touches itself at two points; any plane section through these points gives a quartic curve which touches itself twice, and therefore necessarily consists of two conics having double contact.

The equation of such a quartic surface is $\Phi^2 = \alpha\beta\gamma\delta$ where $\alpha = 0$, $\beta = 0$, $\gamma = 0$, $\delta = 0$ are four coaxal planes; the intersection of their axis with $\Phi = 0$ gives the two double points having the above property; they are usually called *tacnodes*.

Consider next the case in which one of the four double points is a point of contact of the plane and the surface; if none of the three remaining points are fixed the surface possesses a double curve of the third order, which, when a twisted cubic, a line and a conic, or three lines, gives rise to a ruled surface, and the section by a tangent plane to a line and a cubic, excepting only in the case in which the three lines are concurrent. The surface then has the equation

$$Aq^2r^2 + Br^2p^2 + Cp^2q^2 + 2pqrs = 0.$$

This surface is known as that of Steiner.

If next *one* of the double points is fixed, the surface must have a double conic and one node. Its equation is then (Art. 38)

$$\Phi^2 = 4p^2\Psi,$$

where $\Psi = 0$ is a cone whose vertex lies upon $\Phi = 0$. This cone touches the surface along the curve (Φ, Ψ), the tangent planes of Ψ thus meet the surface in pairs of conics.

Lastly let two of the points be points of contact, the surface has a double conic, its equation may be written

$$(\Phi + 2\lambda p^2)^2 = 4p^2(\Psi + \lambda\Phi + \lambda^2 p^2).$$

We may determine λ in five ways so that

$$\Psi + \lambda\Phi + \lambda^2 p^2 = 0$$

is a cone V (Art. 22); V has double contact with the surface; hence the tangent planes of five cones are bitangent planes of the surface and meet it in pairs of conics.

In the case of ruled quartic surfaces, the bitangent planes contain two generators and therefore meet the surface also in a conic.

85. The quartic surfaces which have respectively a double conic and a double line are discussed in Chapters III—VI, that which has three concurrent double lines (Steiner's surface) in the course of this chapter.

To return to the surface $\Phi^2 = \alpha\beta\gamma\delta$, where $\alpha, \beta, \gamma, \delta$ are coaxal planes; this surface has been shown by Nöther* to be birationally transformable into a cubic cone. For if the axis of the planes $\alpha, \beta, \gamma, \delta$ be the line $z=0$, $w=0$, and the planes $x=0$, $y=0$ are the tangent planes to Φ where the line (z, w) meets Φ, then Φ may be written $xy + (z, w \between a)^2$, and the surface becomes

$$\{xy + (z, w \between a)^2\}^2 = (z, w \between b)^4.$$

Choose as new variable w one of the factors of $(z, w \between b)^4$, the quartic surface becomes $\{xy + (z, w \between a)^2\}^2 = w\,(z, w \between b)^3$; then by aid of the transformation

$$x : y : z : w = x'w' - (z', w' \between a)^2 : y'^2 : y'z' : y'w',$$
$$x' : y' : z' : w' = xy + (z, w \between a)^2 : yw : zw : w^2,$$

we obtain $x'^2 w' = (z', w' \between b)^3$; a non-singular cubic cone.

The system of conics on the surface may be represented as follows†:

Writing the equation of the surface in the form

$$\Phi^2 = (z, w \between a)^4 + \left\{\frac{(z, w \between b)^2}{2}\right\}^2,$$

then if
$$(z, w \between a)^4 = a_0 z^4 + a_1 z^3 w + a_2 z^2 w^2 + a_3 z w^3 + a_4 w^4,$$
$$(z, w \between b)^2 = b_0 z^2 + b_1 z w + b_2 w^2,$$

the system of conics is

$$z + \lambda w = 0, \quad z^2 + \mu\left(\Phi - \frac{b_0 z^2 + b_1 z w + b_2 w^2}{2}\right) = 0,$$

with the condition

$$(a_0\mu^2 + b_0\mu - 1)\,\lambda^4 - (a_1\mu^2 + b_1\mu)\,\lambda^3 + (a_2\mu^2 + b_2\mu)\,\lambda^2 - a_3\mu^2\lambda + a_4\mu^2 = 0.$$

From these equations it is seen that the conics of the surface can be arranged in sets of four, lying on the same quadric. By

* *Eindeutige Raumtransformation*, Math. Ann. III.

† Sisam, *Concerning systems of conics lying on cubic, quartic and quintic surfaces*, American Jour. Math. 1908.

suitable choice of b_0, b_1, b_2 we may replace the first two of the preceding equations by

$$z + \lambda w = 0, \quad z^2 + \mu xy = 0,$$

and we obtain the result: *the quadric cone whose vertex is any point of the line $x = y = 0$* and which contains any conic of the system will meet the surface in four conics.*

To the foregoing surfaces described by Kummer we must add the surface whose equation is

$$\{xw + f(y, z, w)\}^2 = (z, w \between a)^4.$$

This may be regarded as a *geometrically*† limiting case of the last surface when the two tacnodes coincide in the point (y, z, w). Its sections by planes through (z, w) consist of pairs of conics.

The equation of the surface may be written

$$(xw - y^2)^2 + 2(xw - y^2)(z, w \between a)^2 + z(z, w \between b)^3 = 0;$$

and by application of the transformation

$$x : y : z : w = y'^2 + z'x' : y'w' : z'w' : w'^2,$$
$$x' : y' : z' : w' = xw - y^2 : yz : z^2 : zw$$

the surface is transformed into the cubic cone

$$z'x'^2 + 2x'(z', w' \between a)^2 + (z', w' \between b)^3 = 0‡.$$

86. Steiner's surface.

The surface of the third class with four tropes was first investigated by Steiner §. In accordance with this definition we may take as its equation in plane-coordinates

$$\frac{1}{u_1} + \frac{1}{u_2} + \frac{1}{u_3} + \frac{1}{u_4} = 0.$$

* Since $z = 0$, $w = 0$ are any two planes through the given line $z = 0$, $w = 0$.

† It cannot be derived from it by giving any particular values to the constants. See Berry, *On quartic surfaces which admit of integrals of the first kind of total differentials*, Camb. Phil. Soc. Trans. 1899.

‡ This quartic surface is discussed by de Franchis, *Le superficie irrazionali di quarto ordine di genere geometrico-superficiale nullo*, Rend. Circ. Mat. di Palermo, XIV. It is there shown that the irrational quartic surfaces for which p_g (the geometrical genus) is zero are either cones or birationally transformable into cones: they include the two surfaces last discussed, also the ruled quartic surface with two non-intersecting double lines, the surface $\{xw + f(y, z, w)\}^2 = (z, w \between a)^4$ where f contains y only to the first degree (this is a ruled quartic with a tacnodal line), and also two special quartic surfaces. See also Berry, *loc. cit.*

§ The surface is also known as Steiner's Roman Surface.

The equation of the surface in point-coordinates is seen to be

$$\sqrt{x_1}+\sqrt{x_2}+\sqrt{x_3}+\sqrt{x_4}=0.$$

The coordinates of the points of the surface may therefore be expressed in terms of the coordinates η_i of the points of a plane by means of the equations

$$\rho x_1=(-\eta_1+\eta_2+\eta_3)^2,\quad \rho x_2=(\eta_1-\eta_2+\eta_3)^2,$$
$$\rho x_3=(\ \eta_1+\eta_2-\eta_3)^2,\quad \rho x_4=(\eta_1+\eta_2+\eta_3)^2.$$

By changing to a second tetrahedron of reference, *desmic* to the first (Chap. II), i.e. by writing

$$\begin{aligned} y_4 &= \ \ \ x_1+x_2+x_3+x_4,\\ y_1 &= \ \ \ x_1-x_2-x_3+x_4,\\ y_2 &= -x_1+x_2-x_3+x_4,\\ y_3 &= -x_1-x_2+x_3+x_4,\end{aligned}$$

we finally obtain

$$\sigma y_1=2\eta_2\eta_3,\quad \sigma y_2=2\eta_3\eta_1,\quad \sigma y_3=2\eta_1\eta_2,\quad \sigma y_4=\eta_1^2+\eta_2^2+\eta_3^2.$$

This method of representing the surface on a plane was first given by Clebsch*. Eliminating the η_i we obtain as the equation of the surface

$$y_2^2y_3^2+y_3^2y_1^2+y_1^2y_2^2-2y_1y_2y_3y_4=0.$$

This latter form of the equation of the surface shows the existence of a triple point A_4 ($y_1=y_2=y_3=0$), three double lines, and three nodes. The section of the surface by any tangent plane contains four nodes, and hence *breaks up into two conics*; a characteristic property of this surface.

There are no lines on the surface other than the three double lines; for there is clearly no other line passing through A_4, and if there were a line not passing through A_4 then the section by the plane through this line and A_4 would possess a triple point at A_4, i.e. would consist of this line together with three other lines through A_4.

The correspondence between a point y_i of the surface and a point η_i of the η-plane is of the (1, 1) character, the only exceptions to this being for points of the three double lines; to such a point there correspond *two* points of the η-plane, e.g. for a point of the line $y_1=y_2=0$ we have

$$\eta_3=0,\quad \frac{y_4}{y_3}=\frac{\eta_1^2+\eta_2^2}{2\eta_1\eta_2};$$

giving two points on $\eta_3=0$, for which the values of $\frac{\eta_1}{\eta_2}$ are reciprocal.

* *Ueber die Steinersche Fläche*, Crelle, LXVII.

87. Curves on the surface.

To a curve $\phi(\eta)=0$ of order n on the η-plane there corresponds a twisted curve of order $2n$ on the surface, this being the number of points in which $\phi(\eta)$ meets any conic

$$\Sigma\eta_i^2 + 2\Sigma a_{ik}\eta_i\eta_k = 0.$$

To the straight lines of the η-plane there correspond the ∞^2 conics of the surface; to conics in the η-plane correspond twisted quartics which are either of the second species or are nodal and of the first species; this is seen from consideration of the *rank* of such curves, i.e. the number of their tangents which meet any given straight line, e.g. $\alpha_y = \beta_y = 0$, i.e. the number of planes $\alpha_y + \lambda\beta_y = 0$ which touch the curve. Denoting the results of substitution for the y_i in terms of the η_i in α_y, β_y by u and v, we have to determine the number of conics $u+\lambda v$ which touch ϕ, giving the equations

$$\frac{\partial u}{\partial \eta_i} + \lambda\frac{\partial v}{\partial \eta_i} = \sigma\frac{\partial \phi}{\partial \eta_i} \qquad (i=1, 2, 3).$$

The required points of contact are given as the intersections of

$$\phi = 0, \qquad \left|\frac{\partial u}{\partial \eta_i}\ \frac{\partial v}{\partial \eta_i}\ \frac{\partial \phi}{\partial \eta_i}\right| = 0;$$

and are $n(n+1)$ in number if ϕ has no singular points. We thus obtain *six* as the rank of twisted quartics on the surface, which therefore belong to one of the two classes previously mentioned.

88. *The equations*

$$\rho x_i = f_i(\eta_1, \eta_2, \eta_3) \qquad (i=1, 2, 3, 4)$$

determine a Steiner's surface, if the curves $f_i = 0$ *are conics.* This may be seen as follows:

The conics apolar to each of four conics

$$a_\eta^2 = 0,\quad b_\eta^2 = 0,\quad c_\eta^2 = 0,\quad d_\eta^2 = 0 \quad \text{...........(1)},$$

form the pencil

$$u_\alpha^2 + \lambda u_\beta^2 = 0 \quad \text{......................(2)},$$

the members of which are all inscribed in the same quadrilateral; if p_η is one side of this quadrilateral then $p_\eta^2 = 0$ is apolar to u_α^2 and u_β^2, and therefore belongs to the linear system (1), or

$$a_\eta^2 + \rho b_\eta^2 + \sigma c_\eta^2 + \tau d_\eta^2 = 0;$$

hence this system includes the squares of four lines; by proper selection of the triangle of reference these lines may be represented by the equations

$$\eta_1+\eta_2+\eta_3=0,\quad -\eta_1+\eta_2+\eta_3=0,\quad \eta_1-\eta_2+\eta_3=0,\quad \eta_1+\eta_2-\eta_3=0,$$

whence we again arrive at the equations connecting a point x of Steiner's surface and a point η of the plane.

The conics of the pencil (2) *are the images of the asymptotic lines of the surface**; this may be seen as follows:

The pairs of tangents drawn from each point P of the plane η to the conics (2) form an involution; if p, p' are the double lines of this involution, then since they are harmonic with each pair of tangents, the line-pair pp' belongs to the system (1), hence its image is a pair of coplanar conics on the surface. These conics intersect on each of the three double lines, and their fourth intersection Q corresponds to P. Moreover, since p, p' are the tangents at P to the two conics of (2) which pass through P, it follows that the line-elements of the asymptotic lines at Q correspond to the line-elements at P of the pair of conics belonging to (2) which pass through P. This being true of every pair of corresponding points Q, P the result follows as stated above†.

* See Cotty, *Sur les surfaces de Steiner*, Nouv. Ann. 1908; Lacour, Nouv. Ann. de Math. 1898.

† Analytically we may proceed as follows: let

$$(\eta_1+m\eta_2+n\eta_3)\left(\eta_1+\frac{1}{m}\eta_2+\frac{1}{n}\eta_3\right)\equiv u\,.\,v$$

be a line-pair belonging to the system $\sum_1^3\eta_i^2+2\Sigma a_{ik}\eta_i\eta_k=0$; if $u+du$ be the line consecutive to u passing through the point on u consecutive to (u, v) we have

$$\Sigma\,(u_i+du_i)\,(\eta_i+d\eta_i)=0,$$

and hence since $\Sigma u_i d\eta_i=0$, it follows that $\Sigma\eta_i du_i=0$ so that the lines u, v, du are concurrent, i.e.

$$v_i=\sigma u_i+\rho du_i.$$

From comparison with the values of u_i, v_i given above we have

$$\frac{dm}{dn}=\frac{m-\dfrac{1}{m}}{n-\dfrac{1}{n}}.$$

This gives $$m^2-1=k\,(n^2-1),$$

k being an arbitrary constant. Hence the image of an asymptotic line is a conic touching the four lines

$$m=\pm 1,\quad n=\pm 1.$$

We obtain sub-cases of Steiner's surface when the conics u_α^2, u_β^2 of (2) are related in certain ways, the four sub-cases which arise are the following:

(i) When two common tangents of the pencil (2) coincide, i.e. the conics of the pencil touch two lines and touch a third line at a fixed point;

(ii) three common tangents coincide, i.e. the conics osculate at a fixed point and also touch a fixed line;

(iii) the conics touch two lines where a third line meets them;

(iv) the conics have four consecutive points common.

The cases (iii) and (iv) lead to cubic surfaces.

In case (i) take the intersection of the two lines as vertex A_1 of the triangle of reference, the fixed point as A_3 and the fourth harmonic to A_1A_3 and the two lines as the third side, the equation of the pencil is then

$$2\lambda uw + w^2 - v^2 = 0.$$

The conics apolar to this pencil are

$$A\eta_1^2 + B(\eta_2^2 + \eta_3^2) + 2C\eta_1\eta_2 + 2D\eta_2\eta_3 = 0.$$

The equations connecting a point x of the quartic surface with the point η of the plane are therefore

$$\rho x_1 = \eta_1^2, \quad \rho x_2 = \eta_2^2 + \eta_3^2, \quad \rho x_3 = 2\eta_1\eta_2, \quad \rho x_4 = 2\eta_2\eta_3;$$

giving as the equation of the surface

$$x_3^4 - 4x_1x_2x_3^2 + 4x_1^2x_4^2 = 0.$$

The surface has a triple point through which there pass two double lines, along one of which one sheet of the surface touches the plane x_1. The surface has two tropes $x_2 = \pm x_4$; the plane x_1 meets the surface in the line (x_1, x_3) alone, this line is thus torsal.

In case (ii) the conics (2) are

$$v^2 - 2uw + 2\lambda vw = 0,$$

giving as the apolar conics

$$A\eta_1^2 + B(\eta_2^2 + 2\eta_1\eta_3) + C\eta_3^2 + 2D\eta_1\eta_2 = 0.$$

The connecting equations are here

$$\rho x_1 = \eta_1^2, \quad \rho x_2 = \eta_2^2 + 2\eta_1\eta_3, \quad \rho x_3 = \eta_3^2, \quad \rho x_4 = 2\eta_1\eta_2.$$

The equation of the surface is

$$(4x_1x_2 - x_4^2)^2 - 64x_1^3x_3 = 0.$$

The surface has one trope, the plane x_3. Along the double line $x_1 = x_4 = 0$, the surface touches the plane $x_1 = 0$.

As previously stated, when the conics of the pencil (2) touch two lines at given points, or have four consecutive points in common, we obtain a cubic surface: this is easily seen, by application of the present method.

89. Modes of origin of the surface.

The connection of the previous mode of representation of the surface and the method of treatment of the surface by Reye* by pure geometry is shown as follows:

If we have any quadric transformation

$$\rho x_i = f_i(\alpha_1, \alpha_2, \alpha_3, \alpha_4), \qquad (i = 1, 2, 3, 4),$$

it has been seen that the locus of the point x as the point α describes a plane is a Steiner surface.

Now the preceding equations place two spaces Σ, Σ_1 in correspondence, so that to each set of quadrics $\sum_1^4 \lambda_i f_i = 0$ in Σ there corresponds a plane in Σ_1, to each pencil of quadrics in Σ a pencil of planes in Σ_1, and to each set of eight associated points in Σ a point in Σ_1.

From the foregoing we deduce also the following result: *Steiner's surface and the cubic polar of a plane with reference to a general cubic surface are reciprocal;* for if $U = 0$ is any cubic surface, from the equations

$$\rho u_i = \frac{\partial U}{\partial x_i}, \qquad (i = 1, 2, 3, 4),$$

we deduce by aid of the preceding that as x describes a plane, u envelops a surface which is the reciprocal of a Steiner surface, and the u_i are the coordinates of the polar plane of x with regard to U.

The following method of derivation of the surface is due to Sturm†. Having given a pencil of quadrics and a pencil of planes projective to it, the locus of intersection of a plane and quadric which are in correspondence is a general cubic surface. If we make the further assumptions that the axis of the planes touches two of the cones contained in the pencil of quadrics, and

* *Geom. der Lage.*

† R. Sturm, *Ueber die Römische Fläche von Steiner*, Math. Ann. III.

that each plane through the axis and the vertex of one of these two cones corresponds to that cone, the surface becomes the general cubic surface with four nodes. For these conditions are seen to be satisfied by assuming as equations of the two pencils

$$V_2 + \lambda V_1 = 0, \quad x_2 + \lambda x_1 = 0,$$

where V_1, V_2 are the cones considered and where

$$V_1 \equiv \alpha^2 + x_1\beta, \quad V_2 \equiv \gamma^2 + x_2\delta.$$

The surface is therefore

$$x_1(\gamma^2 + x_2\delta) - x_2(\alpha^2 + x_1\beta) = 0,$$

which has the four nodes

$$\alpha = x_1 = \gamma^2 + x_2(\delta - \beta) = 0, \quad \gamma = x_2 = \alpha^2 - x_1(\delta - \beta) = 0.$$

Reciprocating, it follows that if a pencil of surfaces of the second order is projectively related to a row of points on a straight line, the envelope of the tangent cones drawn from the various points of the line to the corresponding quadrics is a surface of the third class with four tropes, provided that the line meets one conic c^2 belonging to the system in a point A and one conic c'^2 in a point A', and so that A corresponds to c^2 and A' to c'^2.

Curves on the surface.

The following theorem regarding Steiner's surface is specially noticeable:

Every algebraic curve on the surface is of even order.*

Let a curve c^m of order m on the Steiner surface S_4 pass p times through the triple point and let r, r', r'' branches respectively touch the double lines of S_4 at the triple point, and if p_1, p_2, p_3 other branches respectively touch the three tangent planes at the triple point, then

$$p = r + r' + r'' + p_1 + p_2 + p_3.$$

Denoting the double lines by a, a', a'' it is seen that in the plane (a, a') $p + r + r' + p_1$ points of c^m lie at the triple point, the other $m - p - r - r' - p_1$ points of intersection of S_4 with this plane must therefore lie on a and a', suppose that q lie on a, and hence $q' = m - p - r - r' - p_1 - q$ lie on a'. Similarly on a'' there lie $q'' = m - p - r - r'' - p_2 - q$ points. The cone which projects c^m

* See Sturm, *loc. cit.*

from the triple point is of order $m-p$ and has a as $(q+r)$-fold line, a' as $(q'+r')$-fold line, a'' as $(q''+r'')$-fold line; these lines being double on S_4 count $2(q+r+q'+r'+q''+r'')$ times in the intersection of the cone and S_4, hence

$$4(m-p)=m+2\{q+r+q'+r'+q''+r''\},$$

hence from the preceding

$$m=2\{r+p_1+p_2+q\}.$$

If the curve c^m does not pass through the triple point then $p=0$, and hence

$$p_1=p_2=p_3=r=r'=r''=0,$$

and we have $m=2q$.

Thus a curve of order $2n$ which does not pass through the triple point meets each double line in n points. Hence a curve of the fourth order meets each double line twice; if these points of intersection coincide we have a quartic curve of the first species with a double point (Art. 87).

90. Quartic curves on the surface.

We now consider further the quartic curves on the surface. Every quartic curve c^4 which does not pass through the triple point meets each double line twice; through c^4 there passes at least one quadric which meets the surface in another quartic curve c'^4 which also meets the double lines twice and necessarily in the same pairs of points, excluding at present the case in which the quadric contains one of the double lines.

If the conic $\Sigma a_{ik}\eta_i\eta_k=0$ represents c^4, the conic which represents c'^4 must be

$$\Sigma\frac{\eta_i^2}{a_{ii}}+2\Sigma\frac{a_{ik}\eta_i\eta_k}{a_{ii}a_{kk}}=0 \qquad (i\neq k).$$

For if P, P' are the points in which c^4 meets a double line, the first conic passes through one of the two points corresponding to P and one of the points corresponding to P', while the conic representing c'^4 passes through the other two points corresponding to P and P' respectively. But if e.g. the double line is that which is represented by $\eta_3=0$, it was seen that the values of η_1/η_2 corresponding to P are reciprocals (Art. 86), and so also for P' hence the form of the second conic follows.

Each of these quartic curves is of the second species, since each has three apparent double points.

If we form the product of the left sides of the equations representing these conics the terms which arise are of the type

$$(\Sigma\eta^2)^2, \quad \eta_i^2\eta_j^2, \quad \eta_i^2\eta_j\eta_k,$$

hence on substitution we obtain the equation of the quadric which contains both c^4 and c'^4 *.

Unless two of the quantities a_{ii} are equal, the conics c^2, c'^2 do not intersect on a side of the triangle of reference. From this it follows that the quadric containing c^4 and c'^4 cannot be a cone. For if possible let such a cone be $AC - B^2 = 0$, where A, B are common tangent planes of the cone and quartic surface at two intersections of c^4, c'^4; on substituting in this equation for the x_i in terms of the η_i we obtain

$$aa' \,.\, bb' - u^2 = 0$$

as the equation of the pair of conics c^2, c'^2; where a, a' and b, b' are tangents to the respective conics at two of their intersections; but from the form of this equation the lines $a \dots b'$ must be bitangents of the quartic curve $c^2 \,.\, c'^2$, which is impossible.

If for the curve c^4 we have $a_{ii} = a_{kk}$, it meets one double line in two coincident points. The quadric cone having its vertex at this point and passing through c^4 will meet the surface in another quartic curve c'^4. These quartics are of the first species †.

Any quadric through two double lines meets the surface also in a quartic curve of this latter variety, having a double point on the third double line. There is a quadruply infinite number of such nodal quartics on the surface.

In the first case the quadric containing a pair of quartic curves touches the surface four times; in the second case the cone touches the surface twice.

Let U and V be two quadrics through a quartic c^4 of the first species; if the conic c^2 corresponds to c^4 and if c'^2, c''^2 be the conics

* This quadric is found to be $4Z^2 + \Sigma A_{ik} X_i X_k = 0$, where

$$Z \equiv x_4 + \frac{1}{2}\left\{a_{12}\left(\frac{1}{a_{11}} + \frac{1}{a_{22}}\right)x_3 + a_{13}\left(\frac{1}{a_{11}} + \frac{1}{a_{33}}\right)x_2 + a_{23}\left(\frac{1}{a_{22}} + \frac{1}{a_{33}}\right)x_1\right\},$$

the A_{ik} are the minors of the discriminant of $\Sigma a_{ik}\eta_i\eta_k$, and

$$X_1 = \left(\frac{1}{a_{22}} - \frac{1}{a_{33}}\right)x_1, \quad X_2 = \left(\frac{1}{a_{33}} - \frac{1}{a_{11}}\right)x_2, \quad X_3 = \left(\frac{1}{a_{11}} - \frac{1}{a_{22}}\right)x_3.$$

† For if e.g. $a_{11} = a_{22} = 1$ the curve c^4 will also lie on the quadric

$$x_3(x_4 + a_{12}x_3 + a_{13}x_2 + a_{23}x_1) + \tfrac{1}{2}x_1x_2(a_{33} - 1) = 0.$$

related to the residual intersections of the surface by U and V, we have from above

$$U \equiv c^2 . c'^2, \quad V \equiv c^2 . c''^2,$$

(where in the conics the η_i are replaced by the x_i).

Hence $$U + \lambda V \equiv c^2 (c'^2 + \lambda c''^2),$$

so that the quadri-quartics associated with the pencil of quadrics are thus represented by a pencil of conics *and therefore pass through four fixed points.* This is a characteristic property of these curves.

The conic through one vertex of the triangle of reference

$$u \equiv a_{22}\eta_2^2 + a_{33}\eta_3^2 + 2\Sigma a_{ik}\eta_i\eta_k = 0, \qquad (i \neq k),$$

represents a quartic curve through the triple point; it is clear that if in the product

$$\eta_1 \left(\eta_1 + \frac{2a_{12}}{a_{22}}\eta_2 + \frac{2a_{13}}{a_{33}}\eta_3\right) u$$

we substitute for the η_i in terms of the y_i we obtain an equation containing the latter variables only, and in the second degree.

This quadric hence passes through a double line, a conic of the surface and a curve c^4 through the triple point; this curve is of the *second species,* since one set of generators of the quadric meet c^4 in three points. There exist ∞^4 curves c^4 passing through the triple point.

If the representative conic passes through two vertices of the triangle of reference, i.e. is of the form

$$\eta_3^2 + 2\Sigma a_{ik}\eta_i\eta_k = 0, \qquad (i \neq k),$$

we obtain as a quadric which contains it the cone

$$y_1 y_2 + 2y_3 (a_{12}y_3 + a_{13}y_2 + a_{23}y_1) = 0.$$

These curves are the intersection with the surface of cones through two of the double lines; there are ∞^3 such curves.

Finally, the conics through each vertex of the triangle of reference correspond to the plane sections through the triple point.

91. A mode of origin of Steiner's surface is obtained, as shown by Weierstrass*, from consideration of a well-known property of a quadric. This property is the following: if through any given point A_4 of a quadric Q three mutually perpendicular lines

* Schröter, *Ueber die Steinersche Fläche vierten Grades*, Crelle, LXIV.

are drawn to meet Q again in points L, M, N; the normal at A_4 to Q meets the plane LMN in a fixed point. The theorem may be restated in a general form as follows: if A_4 be joined to the vertices of *any* triangle self-polar to a conic c^2 in a given plane x_4, and the three joining lines meet the quadric again in points L, M, N, then the plane LMN meets the line A_4R in a fixed point S, if R is the pole for c^2 of the trace on x_4 of the tangent plane to Q at A_4.

If c^2 is a member of the set of ∞^2 conics $\alpha^2 = 0$, where

$$\alpha^2 \equiv \eta_1 U + \eta_2 V + \eta_3 W,$$

then by giving all values to the η_i the resulting locus of S is a Steiner surface.

Let the equation of Q be

$$x_4 a_x + \sum_1^3 u_{ik} x_i x_k = 0, \qquad (i \neq k).$$

Also let two vertices A_1, A_2 of the tetrahedron of reference be self-conjugate* for each of the conics U, V, W, and therefore for α^2; let

$$U \equiv \sum_1^3 a_{ik} x_i x_k, \quad V \equiv \sum_1^3 b_{ik} x_i x_k, \quad W \equiv \sum_1^3 c_{ik} x_i x_k.$$

Now if x is the point which forms with A_1 and A_2 a self-polar triangle for α^2, we have

$$x_1 \alpha_{11} + x_3 \alpha_{13} = 0, \quad x_2 \alpha_{22} + x_3 \alpha_{23} = 0 \quad \text{..............(1)},$$

where

$$\alpha_{ik} = \eta_1 a_{ik} + \eta_2 b_{ik} + \eta_3 c_{ik}.$$

Let R, or y, be the pole of a_x for α^2, then

$$\frac{\alpha_{11} y_1 + \alpha_{13} y_3}{a_1} = \frac{\alpha_{22} y_2 + \alpha_{23} y_3}{a_2} = \frac{\alpha_{13} y_1 + \alpha_{23} y_2 + \alpha_{33} y_3}{a_3} \quad \text{......(2)}.$$

It will now be shown that if $A_4 x$ meets Q in x', and $A_4 y$ meets the plane $(A_1 A_2 x')$ in y' (S), then the locus of y' is a Steiner surface.

For, from equations (1) and (2) we obtain the x_i and y_i as quadratic functions of the η_i, thus

$$y_1 : y_2 : y_3 = f_1(\eta) : f_2(\eta) : f_3(\eta);$$

also, it is easy to see from equations (1) and (2) that $\dfrac{a_x}{x_3} = \dfrac{f_3(\eta)}{\alpha_{11}\alpha_{22}}$.

* If A_1, A_2 do not both lie on the section of Q by the plane of c^2, we can replace them by the intersections of A_4A_1, A_4A_2 with Q, and c^2 by its projection on any plane through these two new points.

Again

$$\frac{y_4'}{y_3'} = \frac{x_4'}{x_3'} = -\frac{\sum_1^3 u_{ik} x_i' x_k'}{x_3' a_x'} = -\frac{\sum_1^3 u_{ik} x_i x_k}{x_3 a_x} = \frac{f_4(\eta)}{f_3(\eta)}, \text{ from (1)},$$

where $f_4(\eta)$ is a quadratic function of the η_i. Hence

$$y_1' : y_2' : y_3' : y_4' = f_1(\eta) : f_2(\eta) : f_3(\eta) : f_4(\eta).$$

92. Eckhardt's method.

A method of point-transformation applied by Eckhardt* leads easily to properties of the cubic surface with four nodes, and hence by reciprocation to the surface of the third class with four tropes, which is Steiner's surface. This transformation is the following:

$$x_i y_i = \rho, \qquad (i = 1, 2, 3, 4).$$

By use of this method there corresponds to any given plane

$$\Sigma a_i x_i = 0,$$

the surface

$$\Sigma \frac{a_i}{y_i} = 0,$$

which is a cubic surface having the vertices of the tetrahedron of reference as nodes. The equation of the surface in plane coordinates being

$$\Sigma \sqrt{a_i u_i} = 0,$$

it is seen to be of the fourth class.

The four tangent cones of the surface at the nodes are

$$\frac{a_i}{y_i} - \sum_1^4 \frac{a_k}{y_k} = 0, \qquad (i = 1, 2, 3, 4).$$

They are the tangent cones from the nodes to the quadric

$$\Sigma \frac{y_i^2}{a_i^2} - 2\Sigma \frac{y_i y_k}{a_i a_k} = 0,$$

which touches the edges of the tetrahedron of reference; the intersection of this quadric with the plane

$$\frac{y_1}{a_1} + \frac{y_2}{a_2} - \frac{y_3}{a_3} - \frac{y_4}{a_4} = 0$$

* *Math. Ann.* v.

is a conic c^2 which lies on the cubic surface; for by squaring and subtracting from the equation of the quadric we obtain

$$\frac{y_1y_2}{a_1a_2}+\frac{y_3y_4}{a_3a_4}=0,$$

as a quadric through c^2, and this quadric is easily seen to arise by combining the equation of the plane with that of the cubic surface. Hence the curve of intersection of the quadric and the cubic surface consists of three conics lying respectively in this plane and in two others of similar form.

To a quadric through the vertices of the tetrahedron of reference corresponds a quadric through the same points, and if one quadric is a cone so also is the other (since the discriminant of $\Sigma a_{12}x_1x_2$ is equal to that of $\Sigma a_{34}x_1x_2$). Now *through four points two quadric cones can be drawn to touch a given plane and to have their vertices at a given point of that plane**, we therefore obtain for the cubic surface as the corresponding theorem: *through any point of the surface two quadric cones can be drawn to have their vertices at the point and to touch the surface:* otherwise, the tangent cone to the surface having its vertex at any point of it breaks up into two quadric cones. Reciprocating, we again obtain the result for Steiner's surface that its curve of intersection with any tangent plane consists of two conics.

Again we have the theorem that *eight quadrics can be drawn through any four given points to touch a given quadric along a conic*; for taking the given points as vertices of the tetrahedron of reference and the given quadric as $\Sigma a_{ik}x_ix_k=0$, the latter may be written in the form

$$(\Sigma\sqrt{a_{ii}}x_i)^2+2\Sigma\,(a_{ik}-\sqrt{a_{ii}a_{kk}})\,x_ix_k=0;$$

whence the eight planes of the conics of contact are seen to be

$$\Sigma x_i\sqrt{a_{ii}}=0,$$

taking all combinations of the ambiguities.

If the given quadric consists of two planes, the conics of contact break up into pairs of lines, and hence the tangent quadrics must be cones whose vertices lie upon the line of intersection of the two planes; it follows therefore that *through*

* For the cones which pass through the four points and have their vertices at a given point form the pencil $V_1+\lambda V_2=0$ where V_1 and V_2 are two cones fulfilling these conditions, and the cones of this pencil which touch a given plane are given by a quadratic in λ.

four given points there pass eight quadric cones which touch two given planes; by application of the transformation it is seen that if two cubic surfaces have four nodes in common there are eight quadric cones which touch each surface along two twisted cubics; otherwise, the common tangent developable of two cubic surfaces with four common nodes breaks up into eight quadric cones whose vertices lie upon the curve of intersection of the cubic surfaces. By reciprocation we obtain that if two Steiner's surfaces have four tropes in common, their curve of intersection consists of eight conics whose planes touch the common tangent developable of the two surfaces.

93. Quartic surfaces with a triple point.

The quartic surface with a triple point has been discussed by Rohn*; it belongs to the type known as the *monoid*, i.e. the surface of order n with an $(n-1)$-fold point.

We may take as its equation

$$wu_3 + u_4 = 0,$$

where $u_3 = 0$, $u_4 = 0$ are cones of orders 3 and 4 respectively, having their vertices at the triple point $x = y = z = 0$.

These cones intersect in twelve lines lying on the surface.

These lines meet the plane $w = 0$ in twelve points which we may call *principal points* in the representation of the points of the surface by their projections on this plane from the triple point. This gives a (1, 1) correspondence of points between the surface and the plane, in which, however, all the points of one of the twelve lines are represented by one principal point.

In the general case there is no conic on the surface whose plane does not pass through the triple point A_4; for this would require the quadric cone of vertex A_4 and base the conic, to contain six of the twelve lines of the surface, in order to complete its curve of intersection with the surface. There are sixty-six conics lying in the planes passing through two of the twelve lines; they are represented by the lines joining pairs of principal points.

The quadric cone through five of the twelve lines meets the surface also in a twisted cubic passing through A_4; we obtain 792 such cubics which are represented by conics through five of the principal points. The system of ∞^1 conics through any four

* *Ueber die Flächen vierter Ordnung mit dreifachem Punkte*, Math. Ann. XXIV.

of the principal points represents a system of quartic curves on the surface, each quartic having a node at A_4. Such a quartic must have two apparent double points, since its plane projection from any point must have zero deficiency, being in (1, 1) correspondence with a conic. Hence it is of the first species (a quadri-quartic with one node).

The ∞^1 cubics through eight principal points represent twisted quartics without a node, which pass through A_4 and have the same tangent at that point; they have two apparent double points, their plane projections being in (1, 1) correspondence with non-nodal plane cubics; hence they are of the first species.

Any quartic of the first type and any quartic of the second type lie on the same quadric, provided that the twelve lines with which they are associated are *all different*. For the conic and plane cubic respectively corresponding to them intersect in six points (none of which coincide with principal points), hence if we take that member of the pencil of ∞^1 quadrics through the quartic of the second type which also contains any given point of the first quartic, it will meet the latter in $6+2+1=9$ points, i.e. will contain it.

We also have quartics of the second species, obtained as the intersection with the surface of cubic cones having six of the twelve lines as simple lines and a seventh as double line; they are represented by the plane cubics through six principal points which have a node at a seventh principal point and are 5544 in number. These cubics pass through A_4 and are seen to be of the second species, since they have three apparent double points.

The surface will also possess a line not passing through the double point if three of the twelve lines are coplanar. The maximum number of such lines is nineteen*.

The surface may possess a node D; in that case the line joining it to A_4 must lie on the surface and hence is one of the twelve lines of the surface. Moreover, in this case, two of the twelve lines must *coincide*. For in this case any section through A_4D consists of this line together with a cubic passing through D; we may take A_4D as the line $y=z=0$ and the equation of the surface as

$$w\,\{y\,(ax^2+\ldots)+z\,(bx^2+\ldots)\}+y\,(cx^3+\ldots)+z\,(dx^3+\ldots)=0;$$

* See Rohn, *loc. cit.*

the condition that all such cubics should meet on $y=z=0$ is

$$ad = bc,$$

which is the condition that the curves $u_3=0$, $u_4=0$ should have the same tangent at the point where A_4D meets the plane $w=0$. Hence two of the twelve lines coincide. It is also easily seen that if these curves touch, there is a node at the point

$$y = z = wa + cx = 0.$$

The surfaces with six nodes are of special interest: they are of two types; in the one type the nodes have any position, in the other they lie on a conic.

The equation of a surface of the first type, which has a triple point at A and nodes at six points which we may represent by 1, 2, 3, 4, 5, 6, is of the form

$$KF - \rho P\nabla = 0,$$

where $K=0$ is a quadric cone whose vertex is A and which passes through the points 1 ... 5; $F=0$ is a quadric through the seven points A, 1 ... 6; $P=0$ is the plane through the points A, 4, 5; $\nabla=0$ is the cubic surface with four nodes, viz. at A, 1, 2, 3 respectively, and which passes through the points 4, 5, 6.

There remain three undetermined constants, viz. two in F, together with ρ. It can easily be shown that they can be determined so as to give the required surface. For the conditions that the point 6 should be a node are seen to reduce to the following:

$$\frac{\partial F}{\partial x} : \frac{\partial \nabla}{\partial x} = \frac{\partial F}{\partial y} : \frac{\partial \nabla}{\partial y} = \frac{\partial F}{\partial z} : \frac{\partial \nabla}{\partial z} = \frac{\partial F}{\partial w} : \frac{\partial \nabla}{\partial w} = \rho \frac{P}{K},$$

wherein the coordinates of the point 6 are substituted.

These conditions express that $F=0$, $\nabla=0$ should touch at the point 6, and the two undetermined constants in F are thus found; finally ρ is uniquely determined.

We thus obtain one surface whose equation depends only on the coordinates of its singular points.

A remarkable property of the surface is that the sextic tangent cone, whose vertex is any one of the six nodes, breaks up into two cubic cones, each having a double edge passing through the triple point.

For the equation of the surface may be written

$$x^2 (wA + B) + 2x (wC + D) + wE + F = 0,$$

if the node considered is the point (1000).

The tangent cone whose vertex is this point is then

$$(wC + D)^2 = (wA + B)(wE + F);$$

this is seen to have a fourfold edge passing through A_4. It also has five double edges passing through five nodes. Now *one* cubic cone whose vertex is this node can be drawn to contain these five edges, to have the other edge as double edge, and to contain any other edge of the sextic cone. It therefore intersects the latter cone in $4 \times 2 + 2 \times 5 + 1 = 19$ edges, and hence forms part of it.

The foregoing property is also possessed by the symmetroid (Art. 8). It is easy to see that the surface we have just considered is a special case of the symmetroid. For the latter surface is seen (Chap. IX) to arise when from the equations

$$\alpha \frac{\partial S_1}{\partial x_i} + \beta \frac{\partial S_2}{\partial x_i} + \gamma \frac{\partial S_3}{\partial x_i} + \delta \frac{\partial S_4}{\partial x_i} = 0, \quad (i = 1, 2, 3, 4),$$

expressing that the quadric $\alpha S_1 + \beta S_2 + \gamma S_3 + \delta S_4 = 0$ should be a cone, we eliminate the variable x_i, and regard the α, β, γ, δ as point-coordinates.

If we now take the special case in which S_1 is a plane $a_x = 0$, taken doubly, we obtain as the required surface

$$\begin{vmatrix} \alpha a_1^2 + \beta b_{11} + \dots & \alpha a_2 a_1 + \beta b_{21} + \dots & \dots & \dots \\ \alpha a_1 a_2 + \beta b_{12} + \dots & \alpha a_2^2 + \beta b_{22} + \dots & \dots & \dots \\ \dots & \dots & \dots & \dots \\ \dots & \dots & \dots & \dots \end{vmatrix} = 0.$$

This surface has the point (1000) as triple point, since for this point all the second minors of the determinant vanish.

The surface has a node for such values of $\alpha \dots \delta$ as make

$$\alpha (a_1 x + \dots)^2 + \beta S_2 + \gamma S_3 + \delta S_4 = 0$$

a pair of planes XY; and for such points we have

$$\beta S_2 + \gamma S_3 + \delta S_4 \equiv XY - \alpha a_x^2,$$

i.e. $\beta S_2 + \gamma S_3 + \delta S_4 = 0$ represents a cone whose vertex lies on a_x; and there are six such cones since the vertices of all cones which pass through the eight fixed points $S_1 = S_2 = S_3 = 0$, lie on the sextic curve

$$\beta \frac{\partial S_2}{\partial x_i} + \gamma \frac{\partial S_3}{\partial x_i} + \delta \frac{\partial S_4}{\partial x_i} = 0, \qquad (i = 1, 2, 3, 4).$$

Moreover the preceding surface represents the most general quartic monoid with a *given* triple point and with six nodes.

For the equation of such a surface involves $34-10-6=18$ constants; and the determinant involves thirty-four constants which can be reduced to eighteen on multiplying by an arbitrary determinant of four rows.

The second type of quartic monoid with six nodes is represented by the equation

$$K^2-\rho PF=0,$$

where $K=0$ is a quadric cone, $F=0$ a cubic cone with the same vertex, and $P=0$ any plane.

For this surface has clearly a triple point and has the six nodes given by the equations $P=F=K=0$.

It contains twenty-one constants, the same number as the surface last considered* when the triple point is arbitrary.

* For a full discussion of many special cases of the quartic monoid the reader is referred to the memoir by Rohn recently quoted.

CHAPTER VIII

THE GENERAL THEORY OF RATIONAL QUARTIC SURFACES

94. The quartic surfaces so far considered, with a double curve, have been found to be *rational*, i.e. the coordinates of a point on such a surface are expressible as rational functions of two parameters. Surfaces with a triple point are rational, as is seen by projecting the surface from the triple point on any plane.

We shall now investigate the other types of quartic surfaces with a double point which are rational*. If the double point O be a tacnode, i.e. such that every plane through O meets the surface in a quartic curve having two consecutive double points at O†, the equation of the surface has the form

$$x_4^2 x_1^2 + x_4 x_1 \chi_2(x_1, x_2, x_3) + \chi_4(x_1, x_2, x_3) = 0.$$

Projecting the points x of the surface by lines through O to meet the plane x_4 in points y, we obtain

$$\rho x_1 = y_1, \quad \rho x_2 = y_2, \quad \rho x_3 = y_3, \quad \rho x_4 = \frac{-\chi_2(y) \pm \sqrt{\Omega(y)}}{y_1},$$

where $$\Omega(y) \equiv \{\chi_2(y)\}^2 - 4\chi_4(y).$$

The points x of the surface are thus related to points y of a *double plane*; we now obtain rational expressions of the y_i in terms of new variables z_i which render $\sqrt{\Omega(y)}$ rational in the z_i‡.

The equation of the general quartic curve $\Omega(y) = 0$ may be taken as

$$\begin{vmatrix} u_{11} & u_{12} & u_{13} & u_{14} \\ u_{21} & u_{22} & u_{23} & u_{24} \\ u_{31} & u_{32} & u_{33} & u_{34} \\ u_{41} & u_{42} & u_{43} & u_{44} \end{vmatrix} = 0,$$

where $u_{ik} \equiv u_{ki}$, the u_{ik} being linear functions of the y_i§.

* Nöther, *Ueber die rationalen Flächen vierter Ordnung*, Math. Ann. XXXIII.

† *Selbstberührungspunkt.*

‡ Clebsch, *Ueber Flächenabbildungen*, Math. Ann. III.

§ Hesse, Crelle's *Journal*, XLIX.

A system of cubic curves having six-point contact with Ω is

$$\begin{vmatrix} u_{11} & u_{12} & u_{13} & u_{14} & \alpha_1 \\ u_{21} & u_{22} & u_{23} & u_{24} & \alpha_2 \\ u_{31} & u_{32} & u_{33} & u_{34} & \alpha_3 \\ u_{41} & u_{42} & u_{43} & u_{44} & \alpha_4 \\ \alpha_1 & \alpha_2 & \alpha_3 & \alpha_4 & 0 \end{vmatrix} = 0.$$

Denoting this system by $\phi(y, \alpha)$, there are eight systems of ∞^2 nodal cubics; through every point b_i of the plane there passes one member of each system having a node at b_i. For if $\phi(y, \alpha)$ has a node at b_i, the equations

$$\frac{\partial \phi(b, \alpha)}{\partial b_i} = 0, \qquad (i = 1, 2, 3),$$

give eight sets of values for the α_i.

Again consider the quadrics

$$F(y, X) \equiv \Sigma u_{ik} X_i X_k \equiv y_1 F_1 + y_2 F_2 + y_3 F_3 = 0;$$

if ξ_i be one of the eight points of intersection of $F_1 = 0, F_2 = 0, F_3 = 0$, *the coordinates of the tangent planes to the quadric* $F(b, X)$ *at the eight points* ξ, *form the preceding eight sets of values of the* α_i.

For let quantities η_i, α_i be connected by the equations

$$\alpha_i = u_{i1}(b)\,\eta_1 + u_{i2}(b)\,\eta_2 + u_{i3}(b)\,\eta_3 + u_{i4}(b)\,\eta_4, \quad (i = 1, 2, 3, 4),$$

then it is seen that

$$\phi(b, \alpha) = -F(b, \eta)\,\Omega(b).$$

If now $\eta_i = \xi_i$, the plane α_i touches $F(b, X)$ at the point ξ, and since $F(b, \xi)$ vanishes for all values of the b_i we have

$$\frac{\partial \phi(b, \alpha)}{\partial b_i} = 0, \qquad (i = 1, 2, 3).$$

Thus the eight different sets of ∞^2 nodal cubics belonging to the curves $\phi(y, \alpha)$ are obtained by taking for the quantities α_i the coordinates of the ∞^2 planes through the eight points ξ.

We now select *one* of these eight points ξ, and denote the others by $\xi', \xi'', \ldots$. It is seen that to each quadric $F(y, X)$ there corresponds one point y and conversely; and since there is one quadric F which contains a line s through ξ, therefore to each such line s there corresponds one point y; but each point y determines one quadric F which has two generators s through ξ, i.e. to each point y there correspond two lines s. These two lines coincide if F is a cone, i.e. if we have

$$u_{i1}X_1 + u_{i2}X_2 + u_{i3}X_3 + u_{i4}X_4 = 0, \qquad (i = 1, 2, 3, 4),$$

giving the points y of Ω.

Considering the points z the sections of the sheaf of lines s by the plane which is the field of y, we thus obtain the following relationship between the points y and z; to each point z there corresponds one point y, to each point y there correspond two points z, which coincide when y lies on Ω.

To the quadrics F which touch a given plane α_i, but not at ξ, correspond points y lying on the curve $\phi(y, \alpha)$, and these quadrics give rise to a pencil of lines s lying in α; hence to points z lying on a line p correspond points y lying on the curve $\phi(y, \alpha)$. To points y lying on a line p' correspond points z lying on a curve c of order k; since p' meets ϕ in three points, c must meet p in the corresponding points, i.e. $k = 3$.

Hitherto s has been taken as a line through ξ which does not pass through any of the seven points ξ', ξ'', etc. But if s passes through ξ' then to s will correspond a pencil of quadrics F which determine a line of points y. Denoting by A_i the points in which $\xi\xi'$, $\xi\xi''$, etc. meet the plane of reference, then to each point A_i there corresponds a line, this line meets p' in one point, hence c passes through each of the seven points A_i.

Thus the curves c which correspond to lines in the field of y form a system of ∞^2 cubics through seven fixed points. Such a system is represented by the equation

$$f_1 + \alpha f_2 + \beta f_3 = 0,$$

where f_1, f_2, f_3 are three cubics having seven points in common. The relationship between the points y and z is therefore expressed by the system

$$\sigma y_i = f_i(z), \qquad (i = 1, 2, 3),$$

where the seven points have a general position. Any curve of the system which passes through a given point Q will also pass through a point Q', where Q, Q' correspond to the same point y; the pair of points z which correspond to a point y lie on the same cubic through the seven fixed points.

This transformation rationalizes $\sqrt{\Omega(y)}$; for if $\Delta(z) = 0$ be the curve which corresponds to $\Omega(y) = 0$, since Ω is the locus of points y for which the corresponding pair of points z coincide, i.e. for which curves of the system f_i touch each other, we have

$$\Delta(z) \equiv \begin{vmatrix} \frac{\partial f_1}{\partial z_1} & \frac{\partial f_1}{\partial z_2} & \frac{\partial f_1}{\partial z_3} \\ \frac{\partial f_2}{\partial z_1} & \frac{\partial f_2}{\partial z_2} & \frac{\partial f_2}{\partial z_3} \\ \frac{\partial f_3}{\partial z_1} & \frac{\partial f_3}{\partial z_2} & \frac{\partial f_3}{\partial z_3} \end{vmatrix} = 0;$$

a sextic curve having nodes at the seven points.

When $\sigma^4\Omega(y)$ is expressed in terms of the variables z we obtain an expression of order 12 in the z_i; hence it must be identical with $\{\Delta(z)\}^2$ save as to a constant factor.

The required expression of the surface is therefore

$$\rho' x_1 = f_1(z), \quad \rho' x_2 = f_2(z), \quad \rho' x_3 = f_3(z), \quad \rho' x_4 = \frac{-\psi \pm \kappa\Delta}{f_1(z)},$$

where $\psi \equiv \chi_2(f_1, f_2, f_3)$, and κ is a constant.

Since $$\Omega(y) \equiv \chi_2^2(y) - 4\chi_4(y),$$

we have $$\kappa^2 |\Delta(z)|^2 - [\chi_2\{f(z)\}]^2 = -4\chi_4\{f(z)\},$$

or $$(\kappa\Delta - \chi_2)(\kappa\Delta + \chi_2) \equiv -4\chi_4\{f(z)\}.$$

The plane sections of the surface are represented by two sets of sextic curves, viz.

$$\beta_1 f_1^2 + \beta_2 f_1 f_2 + \beta_3 f_1 f_3 + \beta_4(-\psi \pm \kappa\Delta) = 0.$$

Also $f_1(z) = 0$ meets $\chi_4\{f(z)\} = 0$ in eight points, apart from the seven fixed points, of which four lie upon $\kappa\Delta - \chi = 0$, and four lie upon $\kappa\Delta + \chi = 0$; hence the preceding sextic curves are seen to have the seven fixed points as nodes and to pass through four other fixed points, all of which lie on one cubic curve.

This surface may be referred to as $S_4^{(1)}$.

95. The rational quartic surfaces $S_4^{(2)}$ and $S_4^{(3)}$.

In addition to the rational quartic surface just considered there are two others only, apart from the surfaces which have a double curve or a triple point*. For if the surface is

$$x_4^2 f_2 + 2x_4 f_3 + f_4 = 0,$$

as before its points may be projected from A_4 upon a *double plane*, the plane α_4; if the surface is rational the points y_i of α_4 are such

* Nöther, *Ueber die rat. Flächen vierter Ordnung*, Math. Ann. XXXIII.

rational functions of the coordinates z_i of a simple plane, as rationalize

$$\sqrt{f_3^2(y) - f_2(y) f_4(y)}, \text{ or } \sqrt{\Omega}.$$

In this mapping of the surface upon the plane of the z_i, to plane sections through A_4, or lines in the y-plane, will correspond, in the simple plane, curves of order n, of the same genus as the plane sections, viz. two, and intersecting each other in *two* variable points only. Hence if these curves have in common α_1 points, α_2 double points, ... α_r points of multiplicity r, we have

$$n^2 - 2 = \alpha_1 + 4\alpha_2 + \dots + r^2\alpha_r \quad \text{.................(1)},$$

$$\frac{n(n+3)}{2} - 1 = \alpha_1 + 3\alpha_2 + \dots + \frac{r(r+1)}{2}\alpha_r \quad \text{.........(2)},$$

whence we derive

$$3n = \alpha_1 + 2\alpha_2 + \dots + r\alpha_r \quad \text{.................(3)}.$$

By use of the quadratic transformation

$$z_1 : z_2 : z_3 = z_2'z_3' : z_3'z_1' : z_1'z_2',$$

where the vertices of the triangle of reference are the three multiple points of highest order r, s and t, these curves are transformed into curves of order $2n - r - s - t$, of genus two, and which meet in two variable points. The transformed curves have three corresponding multiple points of orders

$$n - s - t, \quad n - r - t, \quad n - r - s$$

respectively, and other multiple points of the same orders as those of the original curve. Since $r + s + t$ is in all cases* greater

* For from equations (1) and (3) we have

$$3nr - (n^2 - 2) = (r-1)\alpha_1 + 2(r-2)\alpha_2 + \dots + (r-1)\alpha_{r-1}.$$

If $r \geqslant 3$ it is seen that the right-hand side is greater than 2, i.e. $r > \frac{n}{3}$. If $r = 2$ one solution of (1) and (3) is $n = 6$, $\alpha_1 = 2$, $\alpha_2 = 8$; if $n > 6$ there is no solution. Next let r, s and t be not all equal, then if $r = t + \alpha$, $s = t + \beta$, $\alpha \neq 0$, we have

$$n^2 - 2 - r^2 - s^2 = \alpha_1 + 4\alpha_2 + \dots + t^2\alpha_t,$$

$$3n - r - s = \alpha_1 + 2\alpha_2 + \dots + t\alpha_t;$$

hence

$$(3n - r - s)\,t > n^2 - 2 - r^2 - s^2,$$

$$(3n - 2t - \alpha - \beta)\,t > n^2 - 2 - 2t^2 - 2t(\alpha + \beta) - \alpha^2 - \beta^2,$$

whence

$$t > \frac{n^2 - 2 - \alpha^2 - \beta^2}{3n + \alpha + \beta},$$

therefore

$$3t + \alpha + \beta > \frac{3(n^2 - 2 - \alpha^2 - \beta^2)}{3n + \alpha + \beta} + \alpha + \beta > n + 2\,\frac{n(\alpha + \beta) - \alpha^2 - \beta^2 + \alpha\beta - 3}{3n + \alpha + \beta}.$$

If $\beta = 0$ the numerator of the fraction is positive when $n > 4$; if $\beta \neq 0$, since $n > r + s$ it follows that $n > 2t + \alpha + \beta$ and the numerator is positive, i.e.

$$r + s + t > n.$$

than n, by repeated application of the process we finally arrive at curves $c_6(a_1^2 \dots a_8^2 b_1 b_2)$ or at curves $c_4(a^2 b_1 b_2 \dots b_{10})$*. It will now be seen that the curves $c_4(a^2 b_1 \dots b_{10})$ rationalize $\sqrt{\Omega}$ when $\Omega = 0$ is a sextic curve with a quadruple point, and the curves $c_6(a_1^2 \dots a_8^2 b_1 b_2)$ rationalize $\sqrt{\Omega}$ when $\Omega = 0$ is a sextic curve with two consecutive triple points.

The curves $c_4(a^2 b_1 \dots b_{10})$.

By hypothesis these curves form a linear system of ∞^2 curves, hence there is one member of the system which has a node at b_1 (say), this curve cannot be irreducible, for then its genus would be unity and not two as required. Hence it consists of a cubic and a straight line; this requires that the eleven points $ab_1 \dots b_{10}$ should lie on the same cubic. The system of curves consists therefore of linear combinations of the curves $z_1 c$, $z_2 c$ and f, where f is any quartic of the system, c the cubic through the eleven points, and z_1, z_2 any two lines through the point a.

It can be shown that this system rationalizes $\sqrt{\Omega}$ when $\Omega = 0$ is a sextic curve with a quadruple point. For let $\Omega = 0$ be such a sextic with a quadruple point at A_3, $K = 0$ a conic passing through A_3 and having four-point contact with Ω, $L = 0$ a cubic having a node at A_3 and passing through the four points of contact of K and Ω; L thus contains two parameters. The curves $\Omega - \alpha L^2 = 0$ have A_3 as quadruple point and touch K in the previous four points; if now α be so determined that $\Omega - \alpha L^2 = 0$ passes through an additional point of $K = 0$, it must contain K as a factor, i.e. we have

$$\Omega \equiv L^2 - KM,$$

where $M = 0$ is a quartic curve having a triple point at A_3 and touching Ω in six points.

Taking A_3 as the point $y_1 = 0$, $y_2 = 0$, we have

$$K = K_1 y_3 + K_2, \quad L = L_2 y_3 + L_3, \quad M = M_3 y_3 + M_4,$$

where K_1 is linear in y_1, y_2, etc.

Now if
$$\rho y_1 = z_1(z_3^2 \kappa_1 - 2z_3\lambda_2 + \mu_3) = z_1 N,$$
$$\rho y_2 = z_2(z_3^2 \kappa_1 - 2z_3\lambda_2 + \mu_3) = z_2 N,$$
$$\rho y_3 = -(z_3^2 \kappa_2 - 2z_3\lambda_3 + \mu_4),$$

where κ_1, λ_2, ... are the result of substituting z_1, z_2 for y_1, y_2 in K_1,

* Nöther, *Ueber eine Classe von Doppelebenen*, Math. Ann. XXXIII.

$L_2, \ldots$, we have such a transformation; so that to the lines of the y-plane correspond curves $c_4(a^2b_1 \ldots b_{10})$ in the z-plane, the point a being A_3 and the points b_i the other intersections of the curves

$$z_3^2\kappa_1 - 2z_3\lambda_2 + \mu_3 = 0, \quad z_3^2\kappa_2 - 2z_3\lambda_3 + \mu_4 = 0.$$

This transformation rationalizes $\sqrt{\Omega}$, for if $\Delta(z) = 0$ be the curve corresponding to $\Omega = 0$, then since

$$J(z_1N, z_2N, z_3^2\kappa_2 - 2z_3\lambda_3 + \mu_4) \equiv N\Delta(z),$$

it follows that $\Delta(z) = 0$ is a sextic curve.

Moreover, since

$$\Omega(y) \equiv y_3^2f_4 + y_3f_5 + f_6,$$

we have $$\rho^6\Omega(y) = N^4 \times \text{power of } \Delta(z);$$

and the transformation shows that this power is the square; hence

$$\rho^3\sqrt{\Omega(y)} = N^2\Delta(z)^*.$$

The curves $c_6(a_1^2 \ldots a_8^2b_1b_2)$.

As before, since there are ∞^2 curves forming a linear system $f_1 + \alpha f_2 + \beta f_3 = 0$, the system will contain one curve having a node at b_1 (say), i.e. a curve of genus unity, this curve must therefore break up and consist of two cubics of which one passes through the ten points $a_1 \ldots a_8b_1b_2$ and the other through the points $b_1, a_1 \ldots a_8$. If f is the former, and f' any cubic of the pencil through the points $a_1 \ldots a_8$, and ϕ any sextic of the family, the ∞^2 sextics are included in the system

$$f^2 + \alpha ff' + \beta\phi.$$

The transformation effected by means of this system is

$$\rho y_1 = f^2(z), \quad \rho y_2 = f(z)f'(z), \quad \rho y_3 = \phi(z).$$

The curve $\Omega(y)$ is, as before, the locus of points y for which the pairs of points z come into coincidence†. The curve $\Delta(z)$ which corresponds to $\Omega(y)$ is the Jacobian of f, f' and ϕ; it is a curve of order 9 having $a_1 \ldots a_8$ as triple points and not passing through b_1 or b_2. Since any curve of the above system meets $\Delta(z)$ in

$$54 - 48 = 6 \text{ points},$$

any line will meet $\Omega(y)$ in six points, hence $\Omega(y)$ is a sextic

* If $a = \lambda_2\mu_4 - \lambda_3\mu_3$, $2b = \kappa_2\mu_3 - \kappa_1\mu_4$, $c = \lambda_3\kappa_1 - \lambda_2\kappa_2$, we find that

$$\rho^2K = 2N(cz_3 + b), \quad \rho^3L = N^2(cz_3^3 - a), \quad \rho^4M = -2N^3(bz_3^2 + az_3),$$

hence $$\rho^3\sqrt{\Omega} = \rho^3\sqrt{L^2 - KM} = N^2(a + 2bz_3 + cz_3^2).$$

† *Uebergangscurve*, Clebsch.

curve. Since $\Omega(y)$ and $\Delta(z)$ have the same genus, that of the former is seen to be four.

Moreover, the point $y_1 = 0$, $y_2 = 0$ is a triple point on $\Omega(y)$, because the pencil $f + \alpha f' = 0$ meets $\Delta(z)$ in three variable points only; and since $f^2 + \alpha f f' = 0$ meets $\Delta(z)$ in three variable points and fixed points (corresponding to $y_1 = 0$, $y_2 = 0$), f^2 meets Δ in the latter points only, hence the line $y_1 = 0$ touches each of the three branches of $\Omega(y)$ at the triple point and hence meets $\Omega(y)$ only at that point; therefore $\Omega(y)$ has an equation of the form

$$y_1^3 y_3^3 + y_1^2 y_3^2 Q_2 + y_1 y_3 Q_4 + Q_6 = 0,$$

where Q_i is of order i in y_1, y_2*.

Applying the transformation to $\Omega(y)$ we find that

$$\rho^6 \Omega(y) \text{ is equal to } f^6(z) \times \text{some power of } \Delta(z),$$

and this power is seen to be the square, hence

$$\rho^6 \Omega(y) = \{f(z)\}^6 \{\Delta(z)\}^2.$$

The transformation, therefore, rationalizes $\sqrt{\Omega(y)}$.

96. The surfaces $S_4^{(2)}$ and $S_4^{(3)}$.

It remains to determine the surfaces S_4 which arise from the two preceding cases for $\Omega(y)$, i.e. S_4 being

$$x_4^2 f_2(x_1, x_2, x_3) + 2x_4 f_3(x_1, x_2, x_3) + f_4(x_1, x_2, x_3) = 0;$$

the preceding results require that the curve Ω should either have a quadruple point or two consecutive triple points, where

$$\Omega(y) \equiv f_3^2(y) - f_2(y) f_4(y).$$

Now writing

$$f_2 \equiv \alpha x_3^2 + x_3 A_1 + A_2,$$
$$f_3 \equiv \beta x_3^3 + x_3^2 B_1 + x_3 B_2 + B_3,$$
$$f_4 \equiv \gamma x_3^4 + x_3^3 C_1 + x_3^2 C_2 + x_3 C_3 + C_4;$$

(1) If Ω has $x_1 = 0$, $x_2 = 0$ as a quadruple point the following identities must hold:

$$\beta^2 - \alpha\gamma = 0,$$
$$2\beta B_1 - \alpha C_1 - \gamma A_1 \equiv 0,$$
$$B_1^2 + 2\beta B_2 - \alpha C_2 - A_1 C_1 - \gamma A_2 \equiv 0,$$
$$2\beta B_3 + 2B_1 B_2 - \alpha C_3 - A_1 C_2 - A_2 C_1 \equiv 0.$$

* The absence of the term $y_1 y_3^2 y_2^3$ gives rise to two consecutive triple points; thus making the genus of $\Omega(y)$ *four*, as required.

By considering the possible values for α, β, γ we obtain* when $\alpha \neq 0$, $\beta = 0$, $\gamma = 0$, the surface

$$S_4^{(1)} \equiv x_3^2 (x_4 + B_1)^2 + x_3 (x_4 + B_1)(A_1^2 - A_1 x_4 + 2B_2) + x_4^2 A_2 + 2x_4 B_3 + C_4 = 0;$$

a surface having a tacnode in $x_1 = x_2 = x_4 = 0$, i.e. the surface already arrived at.

For $\alpha = \beta = \gamma = 0$ we have surfaces with either a double line or a triple point and therefore excluded.

For $\alpha = \beta = 0$, $\gamma \neq 0$ we obtain either a surface with a triple point or the surface

$$S_4^{(2)} \equiv (x_4 B_1 + x_3^2)^2 + (x_4 B_1 + x_3^2) x_3 C_1 + 2x_4 B_3 + x_3^2 C_2 + x_3 C_3 + C_4 = 0.$$

(2) When Ω has two consecutive triple points, and hence an equation of the form previously given, we obtain the identities

$$\beta^2 - \alpha\gamma = 0,$$
$$2\beta B_1 - \alpha C_1 - \gamma A_1 \equiv 0,$$
$$B_1^2 + 2\beta B_2 - \alpha C_2 - A_1 C_1 - \gamma A_2 \equiv 0,$$
$$2\beta B_3 + 2B_1 B_2 - \alpha C_3 - A_1 C_2 - A_2 C_1 \equiv x_1^3,$$
$$B_2^2 + 2B_1 B_3 - \alpha C_4 - A_1 C_3 - A_2 C_2 \equiv x_1^2 Q_2,$$
$$2B_2 B_3 - A_1 C_4 - A_2 C_3 = x_1 Q_4.$$

By examining the various possible cases we are led to the one surface

$$S_4^{(3)} \equiv x_4^2 x_1^2 + 2x_4 (x_3 x_1 D_1 + B_3) - x_3^3 x_1 + x_3^2 C_2 + x_3 C_3 + C_4 = 0.$$

The surfaces $S_4^{(1)}$, $S_4^{(2)}$, $S_4^{(3)}$ are thus the only rational quartic surfaces apart from such surfaces as have a multiple curve or a triple point.

* See Nöther, *loc. cit.*, p. 152. The reader is referred to this important memoir for details of the mapping of these surfaces on the plane.

CHAPTER IX

DETERMINANT SURFACES

97. The surfaces of the second and third orders may have their equations expressed in the form $\Delta = 0$, where Δ is a determinant having respectively two and three rows, and whose constituents are linear functions of the coordinates. In the case, however, of surfaces of the fourth order, the surface $\Delta = 0$, i.e. the surface whose equation is

$$\begin{vmatrix} p_x & q_x & r_x & s_x \\ p_x' & \cdots & \cdots & \cdots \\ p_x'' & \cdots & \cdots & \cdots \\ p_x''' & \cdots & \cdots & \cdots \end{vmatrix} = 0,$$

where the p_x, etc. are linear functions of the coordinates, depends upon thirty-three constants, one less than the number connected with the general quartic surface. For Δ contains sixty-four constants, of which one may be taken to be unity, and if we multiply Δ by two determinants of four rows, whose elements are arbitrary constants, first by rows and then by columns respectively, we introduce thirty new arbitrary constants; the number of disposable constants contained in Δ is thus seen to be

$$63 - 30 = 33.$$

The surface $\Delta = 0$ may be obtained in two ways as the locus of points common to four collinear systems of space, viz. either by aid of the equations

$$\left.\begin{aligned} \lambda_1 p_x + \lambda_2 q_x + \lambda_3 r_x + \lambda_4 s_x &= 0 \\ \lambda_1 p_x' + \cdots\cdots\cdots\cdots &= 0 \\ \lambda_1 p_x'' + \cdots\cdots\cdots\cdots &= 0 \\ \lambda_1 p_x''' + \cdots\cdots\cdots\cdots &= 0 \end{aligned}\right\} \quad \cdots\cdots\cdots\cdots(1),$$

by elimination of $\lambda_1 \dots \lambda_4$, or similarly by elimination of $\alpha_1 \dots \alpha_4$ from the equations

$$\left.\begin{array}{l} \alpha_1 p_y + \alpha_2 p_y' + \alpha_3 p_y'' + \alpha_4 p_y''' = 0 \\ \alpha_1 q_y + \dots\dots\dots\dots\dots\dots = 0 \\ \alpha_1 r_y + \dots\dots\dots\dots\dots\dots = 0 \\ \alpha_1 s_y + \dots\dots\dots\dots\dots\dots = 0 \end{array}\right\} \dots\dots\dots\dots(2);$$

the resulting surface being in each case

$$| \, pq'r''s''' \, | = 0.$$

If we introduce an additional equation

$$\lambda_1 a + \lambda_2 b + \lambda_3 c + \lambda_4 d = 0,$$

where $a \dots d$ are constants, the equations (1) give rise to four collinear sheaves of planes; the points in which four corresponding planes intersect form a sextic curve lying upon Δ, viz.

$$\left\| \begin{array}{cccc} p_x & q_x & r_x & s_x \\ \dots & \dots & \dots & \dots \\ \dots & \dots & \dots & \dots \\ p_x''' & \dots & \dots & s_x''' \\ a & b & c & d \end{array} \right\| = 0.$$

This set of ∞^3 sextics will be denoted by c_6.

Similarly we obtain the ∞^3 sextics k_6, viz.

$$\left\| \begin{array}{cccc} p_y & p_y' & p_y'' & p_y''' \\ \dots & \dots & \dots & \dots \\ \dots & \dots & \dots & \dots \\ s_y & \dots & \dots & s_y''' \\ A & B & C & D \end{array} \right\| = 0.$$

Any two curves c_6, k_6 lie on the same cubic surface, viz.

$$\left| \begin{array}{ccccc} p_x & \dots\dots & s_x & A \\ p_x' & \dots\dots & s_x' & B \\ p_x'' & \dots\dots & s_x'' & C \\ p_x''' & \dots\dots & s_x''' & D \\ a & b \;\; c & d & 0 \end{array} \right| = 0.$$

Any two curves c_6 intersect in four points, since this is the number of solutions common to the equations (1) and the equations

$$\lambda_1 a + \dots + \lambda_4 d = 0, \quad \lambda_1 a' + \dots + \lambda_4 d' = 0;$$

for eliminating the x_i from equations (1) we obtain what may be regarded as a quartic surface in the λ_i which meets the line given by the last two equations in four points.

Similarly any two curves k_6 meet in four points.

Any two of the preceding cubic surfaces will meet Δ in curves c_6, k_6; c_6', k_6'; hence if m is the number of intersections of c_6 and k_6', we have $2m+8=36$; hence $m=14$.

98. Correspondence of points upon the surface.

The preceding equations (1) and (2) establish a (1, 1) correspondence of points upon the surface; for by aid of the equations

$$\begin{aligned}
&\lambda_1 p_x + \lambda_2 q_x + \lambda_3 r_x + \lambda_4 s_x = 0, \quad && \alpha_1 p_y + \alpha_2 p_y' + \alpha_3 p_y'' + \alpha_4 p_y''' = 0,\\
&\lambda_1 p_x' + \dots\dots\dots\dots = 0, && \alpha_1 q_y + \dots\dots\dots\dots = 0,\\
&\lambda_1 p_x'' + \dots\dots\dots\dots = 0, && \alpha_1 r_y + \dots\dots\dots\dots = 0,\\
&\lambda_1 p_x''' + \dots\dots\dots\dots = 0, && \alpha_1 s_y + \dots\dots\dots\dots = 0;
\end{aligned}$$

and also of the equations

$$\left.\begin{aligned}
\lambda_1 P_1 + \lambda_2 Q_1 + \lambda_3 R_1 + \lambda_4 S_1 &= 0\\
\lambda_1 P_2 + \dots\dots\dots\dots &= 0\\
\lambda_1 P_3 + \dots\dots\dots\dots &= 0\\
\lambda_1 P_4 + \dots\dots\dots\dots &= 0
\end{aligned}\right\} \dots\dots\dots\dots (3);$$

where $P_1 = \alpha_1 p_1 + \alpha_2 p_1' + \alpha_3 p_1'' + \alpha_4 p_1'''$, etc.;

having given any point x of the surface, a set of values of the λ_i are determined and hence one set of values for the α_i, and finally a point y of the surface.

Regarding the λ_i as point-coordinates, and also the α_i, it is seen that by aid of these equations we pass from a point x of Δ to a point λ of a quartic surface Σ, and thence to a point α of a similar surface Σ', and finally to a point y of Δ.

Hence as the point λ describes a plane section of the surface Σ, the point x describes a curve c_6 of Δ, and the point α describes a curve c_6' of Σ'. The point y describes on Δ a curve which, as seen in the next Article, is of the fourteenth order.

If Δ is a symmetrical determinant, i.e. if

$$p' \equiv q, \quad p'' \equiv r, \quad p''' \equiv s, \text{ etc.},$$

the surfaces Σ, Σ' coincide, Δ becomes the surface known as the symmetroid (Art. 8), and Σ the Jacobian of four quadrics.

99. Trisecants of c_6*.

Effecting any linear substitution for the λ_i merely alters the form of the p_x, etc., hence any curve c_6 will be represented by the

* Schur, *Math. Ann.* xx.

curve obtained by taking $\lambda_4 = 0$ as the linear relation connecting the λ_i; this gives the sextic curve

$$\left\| \begin{matrix} p \dots p''' \\ q \dots\dots \\ r \dots r''' \end{matrix} \right\| = 0.$$

Now the three planes

$$\left.\begin{aligned} \alpha_1 p_y + \alpha_2 p_y' + \alpha_3 p_y'' + \alpha_4 p_y''' &= 0 \\ \alpha_1 q_y + \dots\dots\dots\dots\dots\dots &= 0 \\ \alpha_1 r_y + \dots\dots\dots\dots\dots\dots &= 0 \end{aligned}\right\} \dots\dots\dots\dots(4)$$

will be coaxal if equations (3) are satisfied with λ_4 equal to zero; and since c_6 may be written in the form

$$\left\| \begin{matrix} \alpha_1 p + \dots + \alpha_4 p''' & p' & p'' & p''' \\ \alpha_1 q + \dots\dots\dots\dots\dots\dots & & & q''' \\ \alpha_1 r + \dots\dots\dots\dots\dots\dots & & & r''' \end{matrix} \right\| = 0,$$

the axis of the three planes will intersect c_6 in three points, viz. where it meets the cubic surface

$$| p' q'' r''' | = 0.$$

This axis meets Δ in a fourth point y for which, in addition to equation (4), we have the equation

$$\alpha_1 s_y + \dots + \alpha_4 s_y''' = 0.$$

So that any point x of c_6 determines one set of values $(\lambda_1, \lambda_2, \lambda_3, 0)$, and hence one set of values for the α_i which makes the planes (4) coaxal, and therefore one trisecant of c_6; this line meets Δ in a fourth point y, viz. the point which corresponds to x.

As x describes c_6, these trisecants form a ruled surface of the eighth order whose intersection with Δ is c_6 counted thrice, together with a curve of the fourteenth order, the locus of the points y.

For two of the planes (4) being

$$y_1 P_1 + y_2 P_2 + y_3 P_3 + y_4 P_4 = 0,$$
$$y_1 Q_1 + \dots\dots\dots\dots\dots\dots = 0,$$

their intersection will meet any line p_{ik} if

$$p_{12}(P_1 Q_2 - P_2 Q_1) + p_{34}(P_3 Q_4 - P_4 Q_3) + \dots = 0,$$

where the α_i which occur in the P_i, Q_i satisfy the equation

$$\left\| \begin{matrix} P \\ Q \\ R \end{matrix} \right\| = 0.$$

Hence the points α which give the number of trisecants which meet the line p_{ik} are apparently twelve in number, but of these points *four* are those determined by the equations

$$\left\| \begin{matrix} P \\ Q \end{matrix} \right\| = 0,$$

and these points do not in general satisfy the equation

$$p_{12}(P_1R_2 - P_2R_1) + \ldots = 0,$$

and hence must be excluded. The ruled surface is therefore of the eighth order. It meets Δ in c_6, and also in a curve σ whose intersection with any plane $A_y = 0$ is equal to the number of intersections of

$$\left\| \begin{matrix} P \\ Q \\ R \end{matrix} \right\| = 0,$$

and

$$| \, PQSA \, | = 0,$$

excluding again the four points

$$\left\| \begin{matrix} P \\ Q \end{matrix} \right\| = 0;$$

since the four planes

$$\sum_1^4 y_iP_i = 0, \quad \sum_1^4 y_iR_i = 0, \quad \sum_1^4 y_iS_i = 0, \quad \sum_1^4 y_iA_i = 0,$$

in which the coordinates of one of these latter four points α are substituted in the P_i, R_i, S_i, do not in general concur.

Hence the order of σ is $18 - 4 = 14$; and c_6 is a triple curve on the surface formed by the trisecants.

100. The Jacobian of four quadrics and the symmetroid.

The determinant Δ becomes a symmetrical determinant if between its constituents there exist the identities

$$p_x' \equiv q_x, \; p_x'' \equiv r_x, \; \ldots, \; r_x''' \equiv s_x''.$$

The surface Σ' or $| \, P_iQ_iR_iS_i \, | = 0$, is the Jacobian of four quadrics, for it is seen to be a consequence of the above identities that P_i, Q_i, R_i, S_i are the respective partial derivatives of a quadratic expression in the quantities α. The surface Σ' is easily seen to be identical with Σ. By change of notation, replacing λ_i and α_i by

x_i and y_i respectively, the preceding equations (3) may therefore be written, if $P_i = \frac{\partial S_i}{\partial y_1}$, $Q_i = \frac{\partial S_i}{\partial y_2}$, etc.,

$$\sum_{i=1}^{i=4} x_i \frac{\partial S_j}{\partial y_i} = 0, \qquad (j = 1, 2, 3, 4);$$

they express that the polar planes of the point x_i for the four quadrics S_1, S_2, S_3, S_4 meet in the point y, and reciprocally; we have therefore determined on the Jacobian of four quadrics a connection between pairs of its points; they are termed *corresponding* points on the Jacobian. The previous equations (1) and (2) may be then written, replacing x_i, λ_i, y_i, α_i by α_i, x_i, β_i, y_i respectively,

$$\alpha_1 \frac{\partial S_1}{\partial x_j} + \alpha_2 \frac{\partial S_2}{\partial x_j} + \alpha_3 \frac{\partial S_3}{\partial x_j} + \alpha_4 \frac{\partial S_4}{\partial x_j} = 0, \quad (j = 1, 2, 3, 4);$$

$$\beta_1 \frac{\partial S_1}{\partial y_j} + \beta_2 \frac{\partial S_2}{\partial y_j} + \beta_3 \frac{\partial S_3}{\partial y_j} + \beta_4 \frac{\partial S_4}{\partial y_j} = 0, \quad (j = 1, 2, 3, 4).$$

Regarded as arising from these last equations the Jacobian may be defined as the locus of vertices of the cones included in the set of ∞^3 quadrics $\sum_1^4 \alpha_i S_i = 0$.

The surface Δ arising from elimination of the x_i (or y_i) is called the *symmetroid**; its equation may be written in the form

$$\begin{vmatrix} f_{11} & f_{12} & f_{13} & f_{14} \\ f_{12} & f_{22} & f_{23} & f_{24} \\ f_{13} & f_{23} & f_{33} & f_{34} \\ f_{14} & f_{24} & f_{34} & f_{44} \end{vmatrix} = 0,$$

being derived from the last equations, which may be written

$$\sum_{i=1}^{i=4} x_i f_{ji}(\alpha) = 0, \quad \sum_{i=1}^{i=4} y_i f_{ji}(\beta) = 0, \quad (j = 1, 2, 3, 4).$$

It is to be noticed that these equations establish a (1, 1) correspondence between points α, β of the symmetroid through the intervention of a pair of corresponding points, x, y, on the Jacobian.

The Jacobian has ten lines and the symmetroid ten nodes. To show that this is the case, if we write $\Sigma \alpha_i S_i \equiv \Sigma a_{ik} x_i x_k$, the

* Cayley.

condition that $\Sigma\alpha_i S_i = 0$ should be a pair of planes requires a threefold condition between the coefficients a_{ik}, and the number of solutions in the quantities α_i is equal to the number of pairs of planes. Establishing between the a_{ik} six arbitrary linear relations gives a ninefold relation sufficient to determine the quantities a_{ik}. Taking these six relations to express that the quadric should pass through any six given points, the problem is reduced to determining the number of plane-pairs which pass through six given points and this is clearly ten*.

The axis of such a plane-pair clearly lies on the Jacobian, hence this surface contains ten lines.

For such a point α_i the four planes

$$\frac{\partial}{\partial x_j} \sum_i \alpha_i S_i = 0,$$

or

$$\sum_i x_i f_{ji} = 0,$$

are coaxal, hence all the first minors of $| f_{ik} |$ vanish for this point, which is therefore a node on the symmetroid†; hence the symmetroid has ten nodes; to each node α one *line* of J corresponds.

101. Distinctive property of the symmetroid.

The tangent cone of the symmetroid whose vertex is at a node splits up into two cubic cones. For taking a node as the vertex A_1 of the tetrahedron of reference for the α_i, the equations giving the surface may without loss of generality be taken to be

$$\frac{\partial}{\partial x_i} \{\alpha_1 (x_1^2 + x_2^2) + \alpha_2 S_2 + \alpha_3 S_3 + \alpha_4 S_4\} = 0, \quad (i = 1, 2, 3, 4).$$

The equation of the surface is then

$$\begin{vmatrix} \alpha_1 + f_{11} & f_{12} & f_{13} & f_{14} \\ f_{12} & \alpha_1 + f_{22} & f_{23} & f_{24} \\ f_{13} & f_{23} & f_{33} & f_{34} \\ f_{14} & f_{24} & f_{34} & f_{44} \end{vmatrix} = 0,$$

wherein the f_{ik} are linear functions of the coefficients of S_2, S_3, S_4 and the variables α_2, α_3, α_4.

This is, when expanded,

$$\alpha_1^2 \begin{vmatrix} f_{33} & f_{34} \\ f_{34} & f_{44} \end{vmatrix} + \alpha_1 \{F_{11} + F_{22}\} + \Delta = 0,$$

* Cayley.

† Since the tangent plane of the surface at the point is seen to be indeterminate.

where Δ is the determinant $| f_{ik} |$ and F_{ij} is the coefficient of f_{ik} in Δ. The tangent cone from A_1 is therefore

$$(F_{11}+F_{22})^2 = 4\Delta \begin{vmatrix} f_{33} & f_{34} \\ f_{34} & f_{44} \end{vmatrix};$$

but from a known property of determinants

$$F_{11}F_{22} - F_{12}^2 \equiv \Delta \begin{vmatrix} f_{33} & f_{34} \\ f_{34} & f_{44} \end{vmatrix};$$

hence the tangent cone is

$$(F_{11}-F_{22})^2 + 4F_{12}^2 = 0,$$

and thus consists of two cubic cones*.

It was seen that if the tangent cone whose vertex is one node of a ten-nodal quartic surface breaks up into two cubic cones, then the tangent cone for every other node will also break up into two cubic cones (Art. 8).

In forming the Jacobian surface determined by any four quadrics we may suppose these quadrics replaced by any four pairs of planes belonging to the system; and the general Jacobian surface is formed by aid of *any* four pairs of planes. The surface therefore contains twenty-four constants; hence so also does the symmetroid. The number of constants determining the symmetroid is also seen to be twenty-four from the fact that this is the number of arbitrary constants remaining after expressing that the surface has ten nodes.

102. Construction for the tangent plane at any point of the Jacobian of four quadrics.

The vertices of the cones included in the system $\Sigma\alpha_i S_i$ are given by the equations

$$\alpha_1 \frac{\partial S_1}{\partial x_1} + \alpha_2 \frac{\partial S_2}{\partial x_1} + \alpha_3 \frac{\partial S_3}{\partial x_1} + \alpha_4 \frac{\partial S_4}{\partial x_1} = 0,$$

$$\alpha_1 \frac{\partial S_1}{\partial x_2} + \ldots\ldots\ldots\ldots\ldots\ldots\ldots\ldots = 0,$$

$$\ldots\ldots\ldots\ldots\ldots\ldots\ldots\ldots\ldots\ldots\ldots\ldots$$

$$\alpha_1 \frac{\partial S_1}{\partial x_4} + \ldots\ldots\ldots\ldots\ldots\ldots\ldots\ldots = 0.$$

* Cayley, *Coll. Math. Papers*, VII.

Let y be the point corresponding to x on the Jacobian; differentiate these equations and multiply the results respectively by $y_1 \dots y_4$, then by addition we have, since

$$\sum_{i=1}^{i=4} y_i \frac{\partial S_j}{\partial x_i} = 0, \qquad (j = 1, 2, 3, 4),$$

$$\alpha_1 \sum_1^4 y_i d\frac{\partial S_1}{\partial x_i} + \alpha_2 \Sigma y_i d\frac{\partial S_2}{\partial x_i} + \alpha_3 \Sigma y_i d\frac{\partial S_3}{\partial x_i} + \alpha_4 \Sigma y_i d\frac{\partial S_4}{\partial x_i} = 0,$$

which is seen to be the same as

$$\alpha_1 \Sigma dx_i \frac{\partial S_1}{\partial y_i} + \alpha_2 \Sigma dx_i \frac{\partial S_2}{\partial y_i} + \alpha_3 \Sigma dx_i \frac{\partial S_3}{\partial y_i} + \alpha_4 \Sigma dx_i \frac{\partial S_4}{\partial y_i} = 0.$$

But the polar plane of y for the cone of the system whose vertex is x is

$$\alpha_1 \Sigma \xi_i \frac{\partial S_1}{\partial y_i} + \alpha_2 \Sigma \xi_i \frac{\partial S_2}{\partial y_i} + \alpha_3 \Sigma \xi_i \frac{\partial S_3}{\partial y_i} + \alpha_4 \Sigma \xi_i \frac{\partial S_4}{\partial y_i} = 0,$$

it passes through x, and, from the preceding equation, through every point on the Jacobian consecutive to x; it is therefore the tangent plane to the Jacobian at the point x*.

Hence, the tangent plane of the Jacobian at any point P is the polar plane of P', the corresponding point, for the cone of the system of quadrics whose vertex is P.

Two geometrical definitions of the Jacobian of four quadrics have been already obtained: since the line joining two corresponding points is divided harmonically by any quadric of the system, then assuming arbitrarily any six pairs of corresponding points, the surface may also be defined as the locus of vertices of cones which divide harmonically six given segments†. Two other definitions arise as interpretations of the equations

$$-\alpha_1 \frac{\partial S_1}{\partial x_i} = \alpha_2 \frac{\partial S_2}{\partial x_i} + \alpha_3 \frac{\partial S_3}{\partial x_i} + \alpha_4 \frac{\partial S_4}{\partial x_i}, \quad (i = 1, 2, 3, 4),$$

viz. that the surface is the locus of points of contact of quadrics of the system, or that it is the locus of points which have the same polar plane for any two quadrics of the system.

103. Cubic and quartic curves on the Jacobian.

When the point x describes any line of the Jacobian surface, its corresponding point y describes a twisted cubic on the surface:

* See Baker, *Multiply Periodic Functions*, p. 68.

† Cayley.

for let $x_i = a_i + \rho b_i$, and P_a^1, P_a^2, P_a^3 be the polar planes for any point a of three quadrics of the system; the locus of y as given by the preceding equations is derived from

$$P_a^1 + \rho P_b^1 = 0, \quad P_a^2 + \rho P_b^2 = 0, \quad P_a^3 + \rho P_b^3 = 0;$$

giving a twisted cubic: this cubic will not intersect the locus of x but is seen to intersect any other line on the surface twice*. There are ten of these cubics; they are connected with the preceding (1, 1) relationship between points α, β of the symmetroid, which is seen to have exceptional points in that to each node α_i of the symmetroid there corresponds a curve the locus of β_i, which is of the ninth order and has double points at each of the other nodes. For, taking the node as the vertex A_1 of the tetrahedron of reference, to A_1 there corresponds a line on the Jacobian, to this a cubic on the Jacobian, and finally to the latter a curve passing through each of the other nodes twice. To find the order of the curve, the locus of β, we may take its section by the plane $\beta_1 = 0$; the number of points of section is equal to the number of intersections of the cubic curve

$$x_i = a_i + \rho b_i, \quad \sum_{i=1}^{i=4} x_i \frac{\partial S_j}{\partial y_i} = 0, \qquad (j = 2, 3, 4),$$

with the sextic curve

$$\left\| \frac{\partial S_i}{\partial y_1}, \frac{\partial S_i}{\partial y_2}, \frac{\partial S_i}{\partial y_3}, \frac{\partial S_i}{\partial y_4} \right\| = 0, \qquad (i = 2, 3, 4);$$

and these are seen to be the nine points of intersection of this cubic curve with the cubic surface

$$\left| \frac{\partial S_i}{\partial y_j} \right| = 0, \qquad (i, j = 2, 3, 4).$$

Another set of cubic curves on the Jacobian arise as corresponding to plane sections of the symmetroid through three nodes; these curves intersect the three corresponding lines on the Jacobian twice; there are thus 120 cubics of this kind.

To a plane section through two nodes of the symmetroid correspond on the Jacobian two lines and a twisted quartic, intersecting the two lines twice. These quartics may be determined analytically as follows: taking the plane-pairs which respectively

* This is seen by taking the quadric S_1 as the pair of planes intersecting in another line of the surface.

meet on the two lines as uv, $u'v'$ and S_3, S_4 any two quadrics of the system, the Jacobian is derived from the equations

$$\begin{aligned}
\alpha_1 v \quad &+ \alpha_3 \frac{\partial S_3}{\partial u} + \alpha_4 \frac{\partial S_4}{\partial u} = 0,\\
\alpha_1 u \quad &+ \alpha_3 \frac{\partial S_3}{\partial v} + \alpha_4 \frac{\partial S_4}{\partial v} = 0,\\
\alpha_2 v' &+ \alpha_3 \frac{\partial S_3}{\partial u'} + \alpha_4 \frac{\partial S_4}{\partial u'} = 0,\\
\alpha_2 u' &+ \alpha_3 \frac{\partial S_3}{\partial v'} + \alpha_4 \frac{\partial S_4}{\partial v'} = 0.
\end{aligned}$$

If in addition we have the relation $\alpha_4 = k\alpha_3$, then writing

$$v\frac{\partial}{\partial v} - u\frac{\partial}{\partial u} \equiv \delta, \qquad v'\frac{\partial}{\partial v'} - u'\frac{\partial}{\partial u'} \equiv \delta',$$

the foregoing may be written

$$\delta S_3 + k\delta S_4 = 0, \qquad \delta' S_3 + k\delta' S_4 = 0.$$

This gives the Jacobian as the locus of ∞^1 quadri-quartics, each of which twice meets the lines (u, v), (u', v') (and no other line of the Jacobian).

Any quadric through such a quartic meets the Jacobian in another quartic which twice intersects each of the remaining lines of the surface. In this manner we obtain forty-five pairs of systems of quartics on the surface.

104. Sextic curves on the surfaces.

The points y_i of the section of the Jacobian by the plane $a_y = 0$ have as corresponding points x_i, the points of the curve

$$\left\| \frac{\partial S_1}{\partial x_i}, \frac{\partial S_2}{\partial x_i}, \frac{\partial S_3}{\partial x_i}, \frac{\partial S_4}{\partial x_i}, a_i \right\| = 0, \quad (i = 1, 2, 3, 4).$$

This curve has the ten lines of the surface as trisecants*.

The locus of associated points α_i on the symmetroid is a curve of the fourteenth order passing three times through each node†. For the number of points of section of this curve by any plane $b_\alpha = 0$ is the number of intersections of the preceding sextic with the sextic

$$\left\| \frac{\partial S_i}{\partial x_1}, \frac{\partial S_i}{\partial x_2}, \frac{\partial S_i}{\partial x_3}, \frac{\partial S_i}{\partial x_4}, b_i \right\| = 0, \quad (i = 1, 2, 3, 4);$$

* This is easily seen by taking S_1 to be the pair of planes which intersect in one of the ten lines.

† See Art. 99 for the case of c_{14} in the *general* surface Δ.

and the number of intersections of these curves was seen to be fourteen (Art. 97). Since the sextic curves lie on the same cubic surface, the latter sextic does not meet any of the ten lines.

Again the curve $\left\| \frac{\partial S_i}{\partial x_1} \cdots \frac{\partial S_i}{\partial x_4},\ b_i \right\| = 0$

may be represented by the equations (Art. 99)

$$\alpha_2 \frac{\partial S_2}{\partial x_1} + \alpha_3 \frac{\partial S_3}{\partial x_1} + \alpha_4 \frac{\partial S_4}{\partial x_1} = 0,$$

$$\dots\dots\dots\dots\dots\dots\dots\dots$$

$$\dots\dots\dots\dots\dots\dots\dots\dots$$

$$\alpha_2 \frac{\partial S_2}{\partial x_4} + \dots\dots + \alpha_4 \frac{\partial S_4}{\partial x_4} = 0,$$

hence it is the locus of vertices of cones of the system

$$\alpha_2 S_2 + \alpha_3 S_3 + \alpha_4 S_4,$$

i.e. the locus of vertices of cones which pass through eight associated points.

The locus of the points β_i when $a_y = 0$, is the sextic

$$\| f_{i1},\ f_{i2},\ f_{i3},\ f_{i4},\ a_i \| = 0,$$

which passes once through each of the ten nodes. On the symmetroid the curves c_6 and k_6 are of the same kind, each passes through the ten nodes, they therefore intersect in four other points.

105. Additional nodes on the symmetroid.

If S_1, S_2, S_3 and S_4 have a common point, by taking it as a vertex of the tetrahedron of reference we may write

$$a_{11} = b_{11} = c_{11} = d_{11} = 0$$

in the equations of the respective quadrics, so that the highest power of x_1 involved in the Jacobian is the second, hence this point is a node on the surface. Moreover $f_{11} \equiv 0$ in the equation of the symmetroid, so that each term of the equation of this surface contains as factors two of the expressions f_{12}, f_{13} and f_{14}; hence the intersection of these planes is an additional node on the symmetroid.

Similarly if $S_1 \dots S_4$ have two, three or four points in common we have additional nodes arising on the symmetroid. Take the case in which the quadrics have two points in common; if they are the vertices A_1, A_2 of the tetrahedron of reference we have $f_{11} = f_{22} = 0$, and it is seen that the plane $f_{12} = 0$ is a trope of the symmetroid, also that the line joining the two nodes on the Jacobian lies on the surface. Hence if the S_i have k common points ($k = 1, 2, 3, 4$) the Jacobian has $\frac{k(k-1)}{2}$ additional lines and the symmetroid $\frac{k(k-1)}{2}$ tropes.

If $k = 4$ the equation of the symmetroid assumes the form

$$\sqrt{f_{12}f_{34}} + \sqrt{f_{13}f_{24}} + \sqrt{f_{14}f_{23}} = 0.$$

The condition that this surface should have an additional node was seen to be the existence of an identity of the form

$$Af_{12} + A'f_{34} + Bf_{13} + B'f_{24} + Cf_{14} + C'f_{23} \equiv 0,$$

where $AA' = BB' = CC'$ (Art. 12).

This condition may be written in the form

$$A : A' : B : B' : C : C' = c_1c_2 : c_3c_4 : c_1c_3 : c_2c_4 : c_1c_4 : c_2c_3,$$

the c_i being constants.

On reference to the values of the f_{ik}, if we take $S_1 \equiv \Sigma a_{ik}x_ix_k$, $i \neq k$, etc., the preceding identity is seen to lead to the equation

$$a_{12}c_1c_2 + a_{34}c_3c_4 + a_{13}c_1c_3 + a_{24}c_2c_4 + a_{14}c_1c_4 + a_{23}c_2c_3 = 0,$$

with three others obtained by writing respectively b_{ik}, c_{ik}, d_{ik} for a_{ik}; and these equations express that the quadrics $S_1 \dots S_4$ have an additional point c_i in common; hence the Jacobian has an additional node. If the number of common points of the S_i is six, the symmetroid has sixteen nodes and is therefore a Kummer surface; the Jacobian has then twenty-five lines in all, viz. the original ten and the joins of the six additional nodes (Art. 2).

106. Weddle's surface.

The Weddle surface is the locus of vertices of cones of the system $\sum_1^4 a_iS_i = 0$, where $S_1 = 0, \dots, S_4 = 0$ are quadrics having six points in common. Hence the surface is the locus of vertices of cones which pass through six given points. From the definition it is clear that the surface contains the fifteen lines joining the six points in pairs and also the intersections of the ten pairs of planes which can be drawn through the given points. The surface therefore contains twenty-five lines*. Since through each of the six given points five lines of the surface pass, each of these points is a node of the surface. Since the quadrics have six common points, there are three linearly independent quadrics containing the twisted cubic through the six points; through P any point of the surface draw a chord of the cubic meeting it in L and M, the chord will meet the polar plane of P for each of these three quadrics in the same point P', viz. the fourth harmonic to P, L and M; hence the line joining two corresponding points (Art. 100) P and P' of the surface is a chord of the twisted cubic and is cut harmonically by it.

Since any chord of the cubic is cut harmonically by the surface, any tangent to the cubic meets the surface in three consecutive points, and hence the cubic is an asymptotic line of the surface.

107. Parametric representation of the surface.

The coordinates of any point on the twisted cubic may be represented in terms of a parameter θ by the relations

$$x_1 : x_2 : x_3 : x_4 = \theta^3 : \theta^2 : \theta : 1.$$

If A, B are any two points on the twisted cubic having parameters θ, ϕ, then if L, M are the two corresponding points of the surface on A, B their coordinates are given by the relations

$$x_1 : x_2 : x_3 : x_4 = m\theta^3 \pm n\phi^3 : m\theta^2 \pm n\phi^2 : m\theta \pm n\phi : m \pm n\dagger;$$

since L, M divide A, B harmonically.

Let $\Sigma a_{ik} x_i x_k = 0$ be any quadric through the six points, then L, M are conjugate points for this quadric; expressing this fact we at once derive the equation

$$m^2 (a_{11}\theta^6 + 2a_{12}\theta^5 + \ldots) = n^2 (a_{11}\phi^6 + 2a_{12}\phi^5 + \ldots);$$

also if $\theta_1 \ldots \theta_6$ are the parameters of the six points, then

$$a_{11}\theta_1^6 + 2a_{12}\theta_1^5 + \ldots = 0,$$

with five similar equations; it follows that

$$\frac{m^2}{n^2} = \frac{f(\phi)}{f(\theta)},$$

where $f(\alpha) \equiv (\alpha - \theta_1)(\alpha - \theta_2)(\alpha - \theta_3)(\alpha - \theta_4)(\alpha - \theta_5)(\alpha - \theta_6)$.

* See Art. 2. † Richmond.

Hence the points of the surface are parametrically represented by the equations

$$x_1 : x_2 : x_3 : x_4$$

$$= \frac{\theta^3}{\sqrt{f(\theta)}} \pm \frac{\phi^3}{\sqrt{f(\phi)}} : \frac{\theta^2}{\sqrt{f(\theta)}} \pm \frac{\phi^2}{\sqrt{f(\phi)}} : \frac{\theta}{\sqrt{f(\theta)}} \pm \frac{\phi}{\sqrt{f(\phi)}} : \frac{1}{\sqrt{f(\theta)}} \pm \frac{1}{\sqrt{f(\phi)}} \text{*}$$

108. Systems of points on the surface†.

If we represent the quantity $(l\theta^3 + m\theta^2 + n\theta + p)^2/f(\theta)$ by $F(\theta)$, then $F(\theta) - F(\phi) = 0$ is the tangential equation of a pair of corresponding points on the surface. Let θ, ϕ, ψ be the parameters of any three points on the twisted cubic; they give rise to three pairs of points

$$\alpha\alpha' \equiv F(\phi) - F(\psi), \quad \beta\beta' \equiv F(\psi) - F(\theta), \quad \gamma\gamma' \equiv F(\theta) - F(\phi)$$

connected by the relation

$$\alpha\alpha' + \beta\beta' + \gamma\gamma' \equiv 0.$$

This shows that the six points lie by threes on four coplanar lines, i.e. are the vertices of a plane quadrilateral. Moreover if $\alpha\alpha'$, $\beta\beta'$ are two pairs of corresponding points in a plane, they are conjugate for all quadrics of the system; hence the remaining two vertices of the complete quadrilateral of which they are vertices are also conjugate, and therefore are corresponding points on the surface. Any plane meets the twisted cubic in three points, showing that there are only three pairs of corresponding points on the surface in any given plane.

If θ, ϕ, ψ, χ are parameters of any four points on the cubic, we obtain six pairs of points

$$\alpha\alpha' \equiv F(\phi) - F(\psi), \quad xx' \equiv F(\theta) - F(\chi),$$
$$\beta\beta' \equiv F(\psi) - F(\theta), \quad yy' \equiv F(\phi) - F(\chi),$$
$$\gamma\gamma' \equiv F(\theta) - F(\phi), \quad zz' \equiv F(\psi) - F(\chi),$$

whence arises the relation

$$\alpha\alpha' xx' + \beta\beta' yy' + \gamma\gamma' zz' \equiv 0,$$

showing that the tetrahedra whose vertices are (α, α', x, x'), (β, β', y, y'), (γ, γ', z, z') respectively form a desmic system (Art. 13).

* For another method of obtaining these equations see Bateman, *Proc. Lond. Math. Soc.*, Series 2, Vol. III. p. 227.

† See Bateman, *loc. cit.*

Conjugate quintic curves on the surface.

Let S, S' be two consecutive points on the twisted cubic through the six nodes, R any other point on this cubic; then the sides of the triangle RSS' will meet the surface again in three pairs of points PQ, $P'Q'$, TT' lying by threes on four lines. Hence PP' and QQ' which are ultimately tangents at P and Q intersect in a point T on the surface, and since the corresponding point T' ultimately coincides with S, the polar planes of S with regard to quadrics through the six nodes meet in the point T which lies upon the tangent at S.

As R moves along the cubic the point T remains fixed, the points P, Q describe the curve of contact of the tangents from T to the surface.

Again if U be the point derived from R in the same way that T was derived from S, and R is fixed while S varies, the points P, Q will describe the curve of contact of the tangent cone from U; TP, TQ are the tangents at P and Q to this curve of contact. Now UP, UQ are generators of this cone; hence PU, PT are conjugate tangents to the surface at P; thus the curves obtained by keeping one point on the cubic fixed form a conjugate system. To find their order we insert the coordinates of P in any plane $\Sigma a_i x_i = 0$; if θ be constant we obtain

$$\frac{(a_1\theta^3 + a_2\theta^2 + a_3\theta + a_4)^2}{f(\theta)} = \frac{(a_1\phi^3 + a_2\phi^2 + a_3\phi + a_4)^2}{f(\phi)},$$

which is a sextic in ϕ, but rejecting the solution $\theta = \phi$ we obtain five as the number of points of intersection with any given plane.

As in the case of the Jacobian of any four quadrics the tangent plane at any point P of the surface is the polar plane of the corresponding point P' for the cone of the system whose vertex is P. It is determined analytically as follows: the plane

$$lx_1 + mx_2 + nx_3 + px_4 = 0$$

will pass through the point (θ, ϕ) on the surface if

$$\frac{l\theta^3 + m\theta^2 + n\theta + p}{\sqrt{f(\theta)}} = \frac{l\phi^3 + m\phi^2 + n\phi + p}{\sqrt{f(\phi)}}.$$

It will pass through the consecutive point $(\theta + \delta\theta, \phi + \delta\phi)$ for all values of $\delta\theta : \delta\phi$ provided that

$$\frac{\partial}{\partial\theta}\left(\frac{l\theta^3+m\theta^2+n\theta+p}{\sqrt{f(\theta)}}\right)=0,$$

$$\frac{\partial}{\partial\phi}\left(\frac{l\phi^3+m\phi^2+n\phi+p}{\sqrt{f(\phi)}}\right)=0,$$

and will then be the tangent plane at (θ, ϕ).

These equations show that if θ has a given value and ϕ varies, the tangent plane always passes through the point

$$x_1 : x_2 : x_3 : x_4 = \frac{\partial}{\partial\theta}\left(\frac{\theta^3}{\sqrt{f(\theta)}}\right) : \frac{\partial}{\partial\theta}\left(\frac{\theta^2}{\sqrt{f(\theta)}}\right) : \frac{\partial}{\partial\theta}\left(\frac{\theta}{\sqrt{f(\theta)}}\right) : \frac{\partial}{\partial\theta}\left(\frac{1}{\sqrt{f(\theta)}}\right).$$

If in the preceding, S is the point θ, then the coordinates just given are those of the point T. It is easily seen that the locus of T is a rational curve of the seventh order.

It follows from the preceding equation of the tangent plane that the equation of Weddle's surface in plane coordinates is obtained by expressing that the equation

$$(l\theta^3+m\theta^2+n\theta+p)^2-kf(\theta)=0$$

should have two pairs of equal roots for some value of k.

The differential equation of the asymptotic lines may be arrived at in the following manner: the tangent plane at (θ, ϕ) will pass through a consecutive point $(\theta+\delta\theta,\ \phi+\delta\phi)$ if

$$\delta\theta^2\frac{\partial^2}{\partial\theta^2}\left\{\frac{l\theta^3+m\theta^2+n\theta+p}{\sqrt{f(\theta)}}\right\}=\delta\phi^2\frac{\partial^2}{\partial\phi^2}\left\{\frac{l\phi^3+m\phi^2+n\phi+p}{\sqrt{f(\phi)}}\right\}.$$

Also, since $(lmnp)$ is a tangent plane, we may write

$$\left\{\frac{lx^3+mx^2+nx+p}{\sqrt{f(x)}}\right\}^2-k\equiv\frac{\lambda(x-\theta)^2(x-\phi)^2(x-\alpha)(x-\beta)}{f(x)}.$$

Differentiate this equation twice with respect to x and then write $x=\theta$; since

$$\frac{\partial}{\partial\theta}\left\{\frac{l\theta^3+m\theta^2+n\theta+p}{\sqrt{f(\theta)}}\right\}=0$$

we obtain

$$2\frac{l\theta^3+m\theta^2+n\theta+p}{\sqrt{f(\theta)}}\frac{\partial^2}{\partial\theta^2}\left\{\frac{l\theta^3+m\theta^2+n\theta+p}{\sqrt{f(\theta)}}\right\}$$
$$=2\lambda\frac{(\theta-\phi)^2(\theta-\alpha)(\theta-\beta)}{f(\theta)}$$

together with a similar equation in ϕ: hence the differential equation of the asymptotic lines becomes

$$\frac{(\theta-\alpha)(\theta-\beta)}{f(\theta)}d\theta^2=\frac{(\phi-\alpha)(\phi-\beta)}{f(\phi)}d\phi^2,$$

where α and β may be regarded as defined in terms of θ and ϕ by the fact that

$$kf(x) + \lambda(x-\theta)^2(x-\phi)^2(x-\alpha)(x-\beta)$$

is a perfect square.

109. Forms of the equation of the surface.

The surface being defined as the locus of the vertices of cones through six given points, let p_{ik} denote the coordinates of the line joining x to the point a, and q_{ik} the coordinates of the line joining x to the point b, the other given points being the vertices of the tetrahedron of reference. Then since the six lines (x, a), (x, b), $(x, A_1) \ldots (x, A_4)$ lie on a quadric cone, the anharmonic ratio of the pencil formed by the planes $(p, A_1) \ldots (p, A_4)$ is equal to the anharmonic ratio of the pencil $(q, A_1) \ldots (q, A_4)$. But these two anharmonic ratios are determined by the ratios

$$p_{12}p_{34} : p_{13}p_{42} : p_{14}p_{23}$$

and

$$q_{12}q_{34} : q_{13}q_{42} : q_{14}q_{23}$$

respectively; hence the equation of the surface is

$$\frac{p_{12}p_{34}}{p_{13}p_{42}} = \frac{q_{12}q_{34}}{q_{13}q_{42}} \quad \ldots\ldots\ldots\ldots\ldots\ldots\ldots\ldots(1)^*.$$

The surface may therefore be defined as follows: if $B_1 \ldots B_6$ are six given points, then the locus of a point P such that the anharmonic ratio of the four planes

$$(PB_1B_3),\quad (PB_1B_4),\quad (PB_1B_5),\quad (PB_1B_6)$$

is equal to that of the four planes

$$(PB_2B_3),\quad (PB_2B_4),\quad (PB_2B_5),\quad (PB_2B_6)$$

is a Weddle surface which has $B_1 \ldots B_6$ as nodes. By von Staudt's theorem, this may also be stated: *if the anharmonic ratio of the four points in which PB_1 meets the faces of the tetrahedron $B_3 \ldots B_6$ is equal to that of the points in which PB_2 meets this tetrahedron, the locus of P is a Weddle surface with $B_1 \ldots B_6$ as nodes.*

The point $\dfrac{a_i b_i}{x_i}$, or x', is seen to lie on the surface; since writing x' for x in the p_{ik}, q_{ik} of equation (1) merely interchanges the right and left sides of the equation (1).

* See also Hierholzer, *Ueber Kegelschnitte im Raume*, Math. Ann. II., and *Ueber eine Fläche vierter Ordnung*, Math. Ann. IV.

The coordinates of the line joining a and x' are seen to be $\frac{a_i a_k}{x_i x_k} q_{ik}$; if this line intersects the line joining b and x, or q, we must have

$$q_{12}q_{34}(a_1a_2x_3x_4 + a_3a_4x_1x_2) + \ldots + \ldots = 0;$$

which we may write in the form

$$F_{12}q_{12}q_{34} + F_{13}q_{13}q_{42} + F_{14}q_{14}q_{23} = 0.$$

Now expressing the p_{ik} in terms of the x_i and a_i in equation (1) we obtain

$$q_{13}q_{42}(F_{13} - F_{14}) - q_{12}q_{34}(F_{14} - F_{12}) = 0,$$

that is $$F_{12}q_{12}q_{34} + F_{13}q_{13}q_{42} + F_{14}q_{14}q_{23} = 0,$$

since $$q_{12}q_{34} + q_{13}q_{42} + q_{14}q_{23} = 0.$$

Hence the line (a, x') intersects the line (b, x). Expressing that these four points are coplanar we obtain as the equation of the surface

$$\begin{vmatrix} \frac{a_1b_1}{x_1} & \frac{a_2b_2}{x_2} & \frac{a_3b_3}{x_3} & \frac{a_4b_4}{x_4} \\ x_1 & x_2 & x_3 & x_4 \\ a_1 & a_2 & a_3 & a_4 \\ b_1 & b_2 & b_3 & b_4 \end{vmatrix} = 0 \quad \ldots\ldots\ldots\ldots\ldots(2).$$

The surface is thus seen to be completely determined when the six nodes are given. It therefore depends upon eighteen constants. The conditions of possessing six nodes and of containing the joins of five of them require a surface to be a Weddle surface; for the number of constants of the surface remaining arbitrary is

$$34 - 6 \times 4 - 10,$$

which is zero.

On the surface there lie two systems of quadri-quartics, viz. those given by the equations

$$p_{12}p_{34} = \lambda p_{13}p_{42}, \quad q_{12}q_{34} = \lambda q_{13}q_{42};$$

and therefore the intersection of two cones passing through four nodes, and having their vertices at a and b respectively; and those given by the equations

$$p_{12}p_{34} = \mu q_{12}q_{34}, \quad p_{13}p_{42} = \mu q_{13}q_{42};$$

which represent two quadrics through four nodes.

The point x' lies on the curve of the first system determined by the point x. The curves of the first system include, as a special case, the line joining the nodes a and b, together with the twisted cubic through the six nodes.

110. Group of thirty-two points on the surface.

It was seen (Art. 109) that the lines (a, x) and (b, x') intersect; denote their point of intersection by X, so that we have

$$\rho a_i + \kappa X_i = x_i, \quad \sigma b_i + \tau X_i = x_i', \quad (i = 1, \ldots 4);$$

and since $x_i x_i' = a_i b_i$, it follows that

$$(\rho a_i + \kappa X_i)(\sigma b_i + \tau X_i) = a_i b_i, \qquad (i = 1, \ldots 4).$$

These last equations express that the point X lies on the surface.

Similarly, the lines (a, x') and (b, x) meet in a point X' on the surface, such that $X_i X_i' = a_i b_i$. This leads to a system of twenty-two points on the surface; viz. the point x, six points such as X, X' on the lines joining the six nodes to x, and fifteen points such as x', viz. *one* in each plane through a pair of nodes and the point x. From any one of these points the remainder may be derived.

Another system of ten points connected with these twenty-two points is obtained as follows: the nodes being $N_1 \ldots N_6$, then denoting the remaining intersection of the line (N_1, x) with the surface by (N_1), and the remaining intersection of the line $\{N_2, (N_1)\}$ with the surface by $(N_1 N_2)$, and of $\{N_3, (N_1 N_2)\}$ by $(N_1 N_2 N_3)$, it will be shown that the points $(N_1 N_2 N_3)$ and $(N_4 N_5 N_6)$ are identical.

To show this we have to find the coordinates of the point in which the line joining a vertex of the tetrahedron of reference meets the surface again. Denoting by $\Delta_{ijk}(x)$ the determinant

$$\begin{vmatrix} x_i & x_j & x_k \\ a_i & a_j & a_k \\ b_i & b_j & b_k \end{vmatrix},$$

it is easy to see that $\Delta_{ijk}(x) : \Delta_{ijk}(x')$ has the same value for each set of suffixes i, j, k; denote its value by $-H(x)$.

The equation to determine the point x_{A_1}, in which the line (x, A_1) meets the surface, is

$$x_1^2 \Delta_{234}(x') + x_1 \{\ldots\} - a_1 b_1 \Delta_{234}(x) = 0;$$

hence the coordinates of x_{A_1} are as the quantities

$$H(x)\frac{a_1b_1}{x_1} : x_2 : x_3 : x_4,$$

i.e.

$$H(x)\,x_1' : x_2 : x_3 : x_4;$$

also since $H(x)=H(x_{A_1})$, the line (x_{A_1}, A_2) meets the surface in the point

$$x_1'H(x) : x_2'H(x) : x_3 : x_4;$$

finally the point (A_1, A_2, A_3) has the coordinates

$$x_1'H(x) : x_2'H(x) : x_3'H(x) : x_4.$$

Moreover the line (A_4, x') meets the surface in a point whose coordinates are

$$x_1' : x_2' : x_3' : x_4H(x').$$

Hence $(A_1, A_2, A_3)\equiv(a, b, A_4)$, since $H(x)\,H(x')=1$. We thus arrive at a *closed* system of thirty-two points on the surface, from any one of which the others may be derived.

111. Cartesian equation of the surface*.

If we take four nodes as being situated at the origin and at the points at infinity of the (Cartesian) axes of coordinates, the others being A and B, the equation of the surface assumes the form

$$\begin{vmatrix} \dfrac{a_1b_1}{x_1} & \dfrac{a_2b_2}{x_2} & \dfrac{a_3b_3}{x_3} & 1 \\ x_1 & \dots & \dots & 1 \\ a_1 & \dots & \dots & 1 \\ b_1 & \dots & \dots & 1 \end{vmatrix} = 0.$$

If the point P, or x, be joined to the points at infinity on the axes and to the origin, these joining lines will, as has been seen, meet the surface again in the points

$$\left\{\frac{a_1b_1}{x_1}H(x),\ x_2,\ x_3\right\},\quad \left\{x_1,\ \frac{a_2b_2}{x_2}H(x),\ x_3\right\},$$

$$\left\{x_1,\ x_2,\ \frac{a_3b_3}{x_3}H(x)\right\},\quad \{x_1/H,\ x_2/H,\ x_3/H\}.$$

Let X be the point in which the line PB meets the surface again, then transferring the origin of coordinates to A, the new coordinates of x, O, and B respectively are

$$x_i - a_i,\quad -a_i,\quad b_i - a_i.$$

* Baker, *Elementary note on the Weddle quartic surface*, Proc. Lond. Math. Soc., Ser. 2, Vol. I. (1903).

Hence, for the former origin, the coordinates of the point in which OX meets the surface again will be

$$\frac{-a_i(b_i-a_i)}{x_i-a_i}+a_i=\frac{a_i(x_i-b_i)}{x_i-a_i};$$

also
$$(X_i-a_i)\left(\frac{x_i}{H}-a_i\right)=a_i(a_i-b_i);$$

from which we find

$$X_i=a_i\left(\frac{x_i}{H}-b_i\right)\Big/\left(\frac{x_i}{H}-a_i\right).$$

Now denoting by $\theta(x)$, $\phi(x)$, $\psi(x)$ the three points derived from x_i by the transformations

$$\theta(x)=\frac{a_ib_i}{x_i},\quad \phi(x)=\frac{a_ib_iH}{x_i},\quad \psi(x)=b_i(x_i-a_i)/(x_i-b_i),$$

it is seen that these points all lie on the surface, and the eight points derived from P by its projection from the nodes O, A and B form the four couples

$$\begin{array}{cc|cc|cc|cc} x & (Oab) & (ba) & (O) & (b) & (aO) & (a) & (bO) \\ x & \phi(x) & \theta(x) & \theta\{\phi(x)\} & \psi\{\phi(x)\} & \psi(x) & \theta\{\psi(\phi(x))\} & \theta\{\psi(x)\} \end{array}$$

or
$$\begin{array}{cc|cc|cc|cc} x & \dfrac{abH}{x} & \dfrac{ab}{x} & \dfrac{x}{H} & X & \dfrac{b(x-a)}{x-b} & \dfrac{ab}{X} & \dfrac{a(x-b)}{x-a} \end{array}.$$

These show, as above, that the point (O, a, b) is identical with (P, Q, R), where the latter point is that obtained by successive projections of x from the points P, Q, R, at infinity on the axes.

112. Geiser's* method of obtaining the surface.

Let $u_1=0, \dots u_6=0$ be the tangential equations of the six given nodes, then the six quadrics $u_1^2=0, \dots u_6^2=0$ are linearly independent and are apolar† to any quadric through the six points. Hence the general equation of a quadric apolar to the system of quadrics through the six points is

$$\sum_1^6 k_iu_i^2=0.$$

* Geiser, Crelle's *Journal*, LXVII. (1867).

† Two quadrics whose equations in point and plane coordinates are $\Sigma a_{ik}x_ix_k=0$, $\Sigma \alpha_{ik}u_iu_k=0$ are said to be apolar when the invariant $\Sigma a_{ik}\alpha_{ik}$ is zero. When the second equation represents two points, it easily follows that they are conjugate.

When this equation represents two points they are conjugate for all quadrics through the six points and are therefore corresponding points on the Weddle surface. We then have an equation of the form

$$LL' \equiv \Sigma k_i u_i^2.$$

Now let $M=0$, $N=0$ be two points which divide the points L, L' harmonically, hence an identity exists of the form

$$M^2 - N^2 \equiv \Sigma k_i u_i^2.$$

It is easily seen that this is the necessary condition in order that any quadric through seven of the eight points $M, N, u_1 \dots u_6$ should pass through the eighth point*; hence every quadric through $u_1 \dots u_6$ and M will pass through N, and every pair of points on LL' which possesses this property divides LL' harmonically. Such a pair of points can only coincide at one of the points L, L'. It is therefore seen that the Weddle surface arises as the locus of points M such that the point conjugate to M in this manner for the six given points u_i coincides with M†. From this point of view the surface has been shown as a linear projection in four dimensions‡; and projectively related to Kummer's surface.

For if we write

$$\xi = yz', \quad \xi' = y'z, \quad \eta = zx', \quad \eta' = z'x, \quad \zeta = xy', \quad \zeta' = x'y,$$

$$\alpha = bc', \quad \alpha' = b'c, \quad \beta = ca', \quad \beta' = c'a, \quad \gamma = ab', \quad \gamma' = a'b,$$

the equation of the general quadric surface through the six nodes of the Weddle surface in Cartesian coordinates (Art. 111), wherein we write a, b, c for a_1, a_2, a_3; 1, 1, 1 for b_1, b_2, b_3; also x, y, z for x_1, x_2, x_3 and x', y', z', a', b', c' for $1-x$, $1-y$, $1-z$, $1-a$, $1-b$, $1-c$, is

$$A(\alpha\xi' - \alpha'\xi) + B(\beta\eta' - \beta'\eta) + C(\gamma\zeta' - \gamma'\zeta) = \frac{\eta - \eta'}{\beta - \beta'} - \frac{\zeta - \zeta'}{\gamma - \gamma'}.$$

Now $$\xi + \eta + \zeta \equiv \xi' + \eta' + \zeta', \quad \xi\eta\zeta \equiv \xi'\eta'\zeta',$$

so that if we interpret $(\xi, \eta, \zeta, \xi', \eta', \zeta')$ as homogeneous point-coordinates in four dimensions, we have a (1, 1) correspondence between the points of our original space and those of a cubic variety in four dimensions. Again those of the quadric surfaces

* Serret, *Géométrie de Direction*, Nouv. Ann. IV. (1865).

† See Bateman, *loc. cit.* p. 228.

‡ Baker, *loc. cit.*, also Hudson, *Kummer's Quartic Surface.*

which pass through a seventh point (x_1, y_1, z_1) or P_1, have as their equation

$$\{A(\alpha\xi' - \alpha'\xi) + B(\beta\eta' - \beta'\eta) + C(\gamma\zeta' - \gamma'\zeta)\}\left(\frac{\eta_1 - \eta_1'}{\beta - \beta'} - \frac{\zeta_1 - \zeta_1'}{\gamma - \gamma'}\right)$$

$$= \left\{\frac{\eta - \eta'}{\beta - \beta'} - \frac{\zeta - \zeta'}{\gamma - \gamma'}\right\}\{A(\alpha\xi_1' - \alpha'\xi_1) + B(\beta\eta_1' - \beta'\eta_1) + C(\gamma\zeta_1' - \gamma'\zeta_1)\};$$

where $\xi_1 = y_1 z_1', \ldots \zeta_1' = x_1' y_1$.

These quadrics all pass through an eighth point (x, y, z) or P, such that

$$\frac{\alpha\xi' - \alpha'\xi}{\alpha\xi_1' - \alpha'\xi_1} = \frac{\beta\eta' - \beta'\eta}{\beta\eta_1' - \beta'\eta_1} = \frac{\gamma\zeta' - \gamma'\zeta}{\gamma\zeta_1' - \gamma'\zeta_1} = \frac{\dfrac{\eta - \eta'}{\beta - \beta'} - \dfrac{\zeta - \zeta'}{\gamma - \gamma'}}{\dfrac{\eta_1 - \eta_1'}{\beta - \beta'} - \dfrac{\zeta_1 - \zeta_1'}{\gamma - \gamma'}}.$$

These three equations determine the four-dimensional line joining $(\alpha\beta, \ldots)$ to $(\xi_1\eta_1, \ldots)$; the remaining intersection of this line with the cubic variety is the point $(\alpha + \lambda\xi_1, \ldots)$ where λ is given by the equation

$$\lambda(\alpha\eta_1\zeta_1 + \beta\zeta_1\xi_1 + \gamma\xi_1\eta_1 - \alpha'\eta_1'\zeta_1' - \beta'\zeta_1'\xi_1' - \gamma'\xi_1'\eta_1') + \xi_1\beta\gamma + \eta_1\gamma\alpha + \zeta_1\alpha\beta - \xi_1'\beta'\gamma' - \eta_1'\gamma'\alpha' - \zeta_1'\alpha'\beta' = 0$$

and corresponds to the eighth intersection P of the quadrics. The points P, P_1 therefore coincide when this straight line touches the cubic variety, this requires that λ should be infinite, so that

$$\frac{\alpha}{\xi_1} + \frac{\beta}{\eta_1} + \frac{\gamma}{\zeta_1} = \frac{\alpha'}{\xi_1'} + \frac{\beta'}{\eta_1'} + \frac{\gamma'}{\zeta_1'};$$

if we insert $\xi_1 = y_1 z_1'$, etc. we obtain another form of equation of the Weddle surface.

This surface thus arises as the interpretation in three dimensions of the twofold of contact of the enveloping cone of a cubic variety in four dimensions, whose vertex is an arbitrary point of the variety. It has been shown* that the intersection of this cone with an arbitrary planar threefold in space of four dimensions is a Kummer surface. We are therefore led to a birational transformation between the Weddle and Kummer surfaces in the form of a projection; the point (x, y, z) of the Weddle surface being birationally connected with the point $(\xi, \eta, \zeta, \xi', \eta', \zeta')$ of the twofold of contact which is projected into a point of the Kummer surface.

* Richmond, *Quarterly Journal*, XXXI., XXXIV.

113. Sextic curves on the surface.

Any quadric through the six nodes meets the Weddle surface W in an octavic curve. This quadric corresponds, as just seen, to a planar threefold and hence the octavic curve to a plane section of the Kummer surface K. If the quadric is a cone its vertex P lies on the Weddle surface W, hence the octavic has a node at P, and therefore the plane section of K is a tangent plane whose point of contact Q corresponds to P. It was also seen that a system of quadrics through the six nodes and one other point corresponds to planar threefolds passing through a line, and hence to plane sections of K through a fixed point A (the intersection of this line with the planar threefold containing K). Hence the sextic curve which is the locus of vertices of cones of the system will therefore correspond to the curve of contact of the tangent cone from A to K.

Since to any two quadrics S, S' through the six nodes there correspond two planes in the space in which K exists, it follows that the vertices of the four cones determined by S and S' correspond to the points of contact of four coaxal tangent planes of K.

When the quadric contains the twisted cubic through the six nodes, the octavic breaks up into this cubic and a quintic curve. If the quadric is a cone these quintics become identical with those discussed in Art. 108*.

Another set of sextic curves is seen to arise as the intersection with W of any cubic surface having nodes at four nodes of W and therefore containing the lines joining those nodes in pairs; for the curve of intersection consists of the six joins of the four nodes and a sextic curve.

114. Expression of the coordinates as double Theta functions.

The coordinates of any point on a Weddle surface can be expressed in terms of double Theta functions†. For the equation of the surface (Art. 109) is satisfied by the substitutions

$$x_1 : x_2 : x_3 : x_4 = c_{01}\theta_{01}\theta_3\theta_{02}\theta_{04} : c_2\theta_2\theta_1\theta_3\theta_{04} : c_{03}\theta_{03}\theta_1\theta_{02}\theta_{04} : c_4\theta_4\theta_1\theta_3\theta_{02},$$

$$a_1 : a_2 : a_3 : a_4 = c_{01}c_0c_{23}c_{34} : c_2c_5c_{12}c_{23} : c_{03}c_0c_{12}c_{14} : c_4c_5c_{14}c_{34},$$

$$b_1 : b_2 : b_3 : b_4 = c_{01}c_5c_{12}c_{14} : c_2c_0c_{14}c_{34} : c_{03}c_5c_{23}c_{34} : c_4c_0c_{12}c_{23},$$

* This is seen at once since the point whose coordinates are $\frac{\theta^3}{\sqrt{f(\theta)}} \pm \frac{\phi^3}{\sqrt{f(\phi)}}$, etc., for $\theta = \text{constant}$, lies on the cone $(x_2 - \theta x_3)^2 = (x_1 - \theta x_2)(x_3 - \theta x_4)$.

† Caspary, *Ueber Thetafunktionen mit zwei Argumenten*, Crelle, xciv.

where c_0 is the result of attributing zero values to the variables in $\theta_0(u)$, etc., as will now be shown.

In the first place we see that the coordinates of the point x' or $\frac{a_i b_i}{x_i}$ are derived from those of x by increasing the argument by $\frac{1}{2}(\tau c+d)$*; since this interchanges θ_1, θ_{01}; θ_2, θ_{02}, etc. Again, as before, let p_{ik} denote the coordinates of the line (a, x) and q_{ik} those of the line (b, x); we find on substitution

$$\frac{p_{12}p_{34}}{p_{13}p_{42}}=\frac{(c_5c_{12}\theta_{01}\theta_{02}-c_0c_{34}\theta_1\theta_2)(c_5c_{34}\theta_{03}\theta_{04}-c_0c_{12}\theta_3\theta_4)\,c_{23}c_{14}}{(c_{12}c_{14}\theta_{01}\theta_3-c_{23}c_{34}\theta_{03}\theta_1)(c_{12}c_{23}\theta_4\theta_{02}-c_{14}c_{34}\theta_2\theta_{04})\,c_0c_5}\;\ldots(1).$$

If q'_{ik} denote the line joining x' to a, it is easily seen that

$$\frac{q_{12}q_{34}}{q_{13}q_{42}}=\frac{q'_{12}q'_{34}}{q'_{13}q'_{42}},$$

and the latter ratio is formed, as stated above, from $\frac{p_{12}p_{34}}{p_{13}p_{42}}$ by increasing the argument of the θ's by $\frac{1}{2}(\tau c+d)$.

It will now be shown that this change does not affect the right side of equation (1). For, as is well known, the determinant

$$\begin{vmatrix} c_5\theta_5 & 0 & 0 & 0 \\ 0 & c_4\theta_4 & -c_{14}\theta_{14} & c_{34}\theta_{34} \\ 0 & c_0\theta_0 & c_{01}\theta_{01} & -c_{03}\theta_{03} \\ 0 & -c_2\theta_2 & c_{12}\theta_{12} & c_{23}\theta_{23} \end{vmatrix}$$

forms an orthogonal matrix, from which we may therefore derive the equations

$$\left.\begin{aligned} c_5\,c_{12}\theta_5\,\theta_{12}-c_0\,c_{34}\theta_0\,\theta_{34} &= c_4\,c_{03}\theta_4\,\theta_{03} \\ c_5\,c_{34}\theta_5\,\theta_{34}-c_0\,c_{12}\theta_0\,\theta_{12} &= c_2\,c_{01}\theta_2\,\theta_{01} \\ c_{12}c_{14}\theta_{12}\theta_{14}-c_{23}c_{34}\theta_{23}\theta_{34} &= -c_2\,c_4\,\theta_2\,\theta_4 \\ c_{12}c_{23}\theta_{12}\theta_{23}-c_{14}c_{34}\theta_{14}\theta_{34} &= c_{01}c_{03}\theta_{01}\theta_{03} \end{aligned}\right\}\;\ldots\ldots\ldots(2).$$

Increasing the arguments in each of these equations by the respective half-periods

$$\tfrac{1}{2}(\tau d+b),\quad \tfrac{1}{2}(\tau a+a),\quad \tfrac{1}{2}(\tau a+b),\quad \tfrac{1}{2}(\tau a+a),$$

the left sides are transformed into the quantities which appear on

* Using the notation of Hudson, *Kummer's Quartic Surface*, p. 178. We compare for convenience Hudson's notation with that just given

$\theta_5\;\theta_{12}\;\theta_0\;\theta_{34}$	$\theta_{23}\,\theta_{13}\,\theta_{14}\,\theta_{24}$	$\theta_4\;\theta_{03}\,\theta_{04}\;\theta_3$	$\theta_{01}\;\theta_{02}\,\theta_1\;\theta_2$
dd ad cd bd	*dc ac cc bc*	*da aa ca ba*	*db ab cb bb.*

The equations for the addition of characteristics being

$$a+a=b+b=c+c=d+d=d;$$

$$b+c=a=a+d,\qquad c+a=b=b+d,\qquad a+b=c=c+d.$$

the right side of equation (1), save as to an exponential factor in each case, and we therefore derive that

$$\frac{p_{12}p_{34}}{p_{13}p_{42}} = -\frac{\theta_{23}\theta_{13}\,.\,\theta_{13}\theta_{14}}{\theta_0\theta_{13}\,.\,\theta_{13}\theta_5} = -\frac{\theta_{23}\theta_{14}}{\theta_0\theta_5},$$

the exponential factors having cancelled out.

Now the increase of the argument of the θ's by $\frac{1}{2}(\tau c + d)$ does not affect the value of $\frac{\theta_{23}\theta_{14}}{\theta_0\theta_5}$, hence

$$\frac{p_{12}p_{34}}{p_{13}p_{42}} = \frac{q_{12}q_{34}}{q_{13}q_{42}},$$

and the point x is seen to lie on the Weddle surface.

By considering the expressions of the p_{ik} and q_{ik} it is seen that the two systems of quadri-quartics on the surface are given by the equations

(i) $\theta_{13} = \lambda\theta_{24}$,

(ii) $\theta_{23}\theta_{14} = \mu\theta_0\theta_5, \quad \theta_{12}\theta_{34} = \nu\theta_0\theta_5, \quad \theta_{12}\theta_{34} = \sigma\theta_{23}\theta_{14}$.

The quantity H is seen to have the value

$$\theta_1\theta_{01}\theta_2\theta_{02}\theta_3\theta_{03}\theta_4\theta_{04} \div c_0c_5c_{12}c_{34}c_{14}c_{23}.$$

The thirty-two points forming a closed system are derived as follows: fifteen of them arise from adding to the argument u for P, the fifteen half-periods $\frac{1}{2}(\tau a + b)$, etc. These are the fifteen points (N_1N_2). The other sixteen points arise in the following manner: since the coordinates of N_1, the other intersection of the line PA_1 with the surface, are

$$Hx_1' : x_2 : x_3 : x_4,$$

it follows, from the above value of H, that the coordinates of N_1 are

$$c_{01}\theta_{03}\theta_2\theta_4 : c_2\theta_2\theta_3\theta_{04} : c_{03}\theta_{03}\theta_{02}\theta_{04} : c_4\theta_4\theta_3\theta_{02},$$

since θ_1 divides out.

The other five points (N_i) and the ten points $(N_1N_2N_3)$ are derived from (N_1) by the addition to its argument of the fifteen half-periods*.

* Weddle's surface is a case of a class of surfaces investigated by Humbert, *Théorie générale des surfaces hyperelliptiques*, Journal de Math., série 4, t. IX. (1893). These surfaces are termed hyperelliptic surfaces, the coordinates of any point are uniform quadruply periodic functions of two parameters; see also Hudson, *Kummer's Quartic Surface*, pp. 182–187.

Baker has shown that the coordinates of any point on the Weddle surface may be expressed as derivatives of a single variable (*Multiply Periodic Functions*, pp. 39, 40, 77).

115. Plane sections of the surface.

The equation of the plane section of a Weddle surface may be simply expressed. Take as triangle of reference the three points in which the given plane meets the twisted cubic through the six nodes. Each side of this triangle meets the curve of section in two vertices and also in two points harmonic with these vertices. Hence we obtain as the equation of the surface

$$a_2x^3y + a_3x^3z + b_3y^3z + b_1y^3x + c_1z^3x + c_2z^3y + 3xyz\,(lx + my + nz) = 0.$$

Also, since the three pairs of points lying on the sides of the triangle of reference lie by threes on four lines (Art. 108), we have the condition

$$a_2b_3c_1 + a_3b_1c_2 = 0.$$

From the last condition we infer that the tangents at the vertices of the triangle of reference are concurrent.

If we form for this quartic the invariants A and B* we find

$$A = -12lmn + 12\,(lb_1c_1 + mc_2a_2 + na_3b_3),$$

$$B = -\begin{vmatrix} l & a_3 & a_2 \\ b_3 & m & b_1 \\ c_2 & c_1 & n \end{vmatrix}^2.$$

Hence for any plane section we have the invariant condition

$$A^2 + 144B = 0.$$

An infinite number of configurations of points can be obtained on the plane section as follows: let the Weddle surface be determined by four quadrics S_1, S_2, S_3, S_4, of which we may suppose the first three to contain the same twisted cubic. Then the section considered contains the following set of twenty-five points, viz. the fifteen points in which the join of two of the nodes $N_1 \ldots N_6$ meets the plane, and the ten points in which the ten lines $(N_1N_2N_3,\ N_4N_5N_6)$ meet the plane. The first set of fifteen points lie by threes on twenty lines, viz. the intersection with the given plane of the planes $(N_1N_2N_3)$, etc.

Now consider the Weddle surfaces formed by aid of S_1, S_2, S_3 and $S_4 + \lambda\alpha^2$, where $\alpha = 0$ is the plane of the section. These surfaces form a pencil whose nodes lie on the same twisted cubic, and all containing the same section lying in $\alpha = 0$; from each surface

* Salmon, *Higher Plane Curves*, 3rd Ed. p. 264.

one configuration, of the kind just mentioned, arises. Hence we have an infinite number of such configurations*.

116. Bauer's surfaces.

If in the foregoing four collinear systems (Art. 97) each plane system reduces to a sheaf, and is such that each plane joining the centres of three sheaves is a self-corresponding plane for three systems, we obtain the surface discussed by Bauer†. The equation of such a surface is accordingly

$$\begin{vmatrix} x_1 - \dfrac{a_x}{a_1} & x_2 & x_3 & x_4 \\ x_1 & x_2 - \dfrac{b_x}{b_2} & x_3 & x_4 \\ x_1 & x_2 & x_3 - \dfrac{c_x}{c_3} & x_4 \\ x_1 & x_2 & x_3 & x_4 - \dfrac{d_x}{d_4} \end{vmatrix} = 0.$$

This equation may also be written in the form

$$\frac{a_1 x_1}{a_x} + \frac{b_2 x_2}{b_x} + \frac{c_3 x_3}{c_x} + \frac{d_4 x_4}{d_x} = 1;$$

wherein a_x, b_x, c_x, d_x are linear functions of the coordinates.

The foregoing equation may also be obtained as follows: a point P (or x) is joined to the vertices of a given tetrahedron Δ (taken as that of reference) and the joining lines PA_1, etc., meet the faces of any other given tetrahedron Δ' (whose faces are $a_x = 0, \dots d_x = 0$) in points $Q_1 \dots Q_4$; then if the points Q_i are coplanar the locus of P is the surface just given. For the coordinates of Q_1 are seen to be $x_1 - \dfrac{a_x}{a_1}$, x_2, x_3, x_4, and expressing that the points Q_i are coplanar, we obtain the foregoing equation.

The second form of equation of the surface shows that the edges of Δ' lie on the surface and also the intersections of corresponding faces of Δ and Δ', as $x_1 = 0$, $a_x = 0$, etc.; the vertices of Δ' are seen to be nodes of the surface. The surface therefore possesses ten lines and four nodes.

Denoting the lines (x_1, a_x), etc. by p_1, etc., if two lines p

* See Morley and Conner, *Plane sections of a Weddle surface*, Amer. Journ. of Math. XXXI.

† Bauer, *Ueber Flächen 4. Ordnung deren geom. Erzeugung sich an 2 Tetraeder knüpft*, Sitz. d. König. Akad. d. Wiss. München, 1888.

intersect, their point of intersection is seen to be a node of the surface; if each line p meets every other line p, then the lines p lie in one plane, say the plane $z=0$, also each edge of Δ meets the corresponding edge of Δ' and the two tetrahedra are *in perspective.* In this case the equation of the surface assumes the form

$$z\{\lambda_1 b_x c_x d_x + \lambda_2 c_x d_x a_x + \lambda_3 d_x a_x b_x + \lambda_4 a_x b_x c_x\} = a_x b_x c_x d_x,$$

where the λ_i are constants. For in this case we may write

$$x_1 = \mu_1 z + \nu_1 a_x, \quad x_2 = \mu_2 z + \nu_2 b_x, \text{ etc.}$$

The surface is the Hessian of the general cubic surface; it has ten nodes of which six lie in $z=0$.

Let now an edge of Δ intersect the edge of Δ' *opposite* to the corresponding edge, e.g. let the line (x_1, x_2) intersect the line (c_x, d_x); in this case it is easily seen that (x_1, x_2) lies on the surface; if this occurs in every case the surface will contain also the six edges of Δ and have the vertices of Δ as nodes*.

Lastly we may assume that both sets of conditions are satisfied, viz. that each edge of one tetrahedron intersects a pair of opposite edges of the other. The tetrahedra are then in *desmic* position (Art. 13).

The equation of this surface, viz.

$$\frac{x_1}{-x_1+x_2+x_3+x_4} + \frac{x_2}{x_1-x_2+x_3+x_4} + \frac{x_3}{x_1+x_2-x_3+x_4} + \frac{x_4}{x_1+x_2+x_3-x_4} + 1 = 0,$$

may be reduced either to the form

$$z(x_1x_2x_3 + x_2x_3x_4 + x_3x_4x_1 + x_4x_1x_2) = 4x_1x_2x_3x_4,$$

where $z=\Sigma x_i$, or to the form

$$\sqrt{Z_1} + \sqrt{Z_2} + \sqrt{Z_3} = 0,$$

where

$$Z_1 = (x_1+x_2)(x_3+x_4), \quad Z_2 = (x_1+x_3)(x_2+x_4),$$
$$Z_3 = (x_1+x_4)(x_2+x_3).$$

* The equation of this surface may be written in the form

$$AZ_1^2 + BZ_2^2 + CZ_3^2 + DZ_1Z_2 + EZ_2Z_3 + FZ_3Z_1 = 0,$$

where the Z_i are pairs of planes through opposite edges of Δ'.

117. Schur's surfaces.

A particular case of the surface Δ arises when the foregoing correspondence between points x, y of Δ is reduced to a collineation*. It has been seen (Art. 99) that as x describes a curve c_6 the corresponding point y describes a curve which is the locus of the fourth intersection with Δ of the trisecants of c_6, but since the points x, y are to be in this case linearly connected, y must also describe a curve of order six, hence the intersection with Δ of the surface formed by the trisecants must include eight of these trisecants $a_1 \dots a_8$, in order to complete the order, 14, of the complete intersection of the surfaces apart from c_6. Similarly every k_6 has eight trisecants $b_1 \dots b_8$ which lie upon Δ.

The lines a and b are distinct and no two lines a intersect each other; similarly no two lines b intersect. For, in this case, to the point x of c_6 which gives rise to a line a_i there corresponds an infinite number of points y, viz. the points of a_i; hence the *four* planes

$$\begin{aligned} \alpha_1 p + \alpha_2 p' + \alpha_3 p'' + \alpha_4 p''' &= 0, \\ \dots\dots\dots\dots\dots\dots\dots &= 0, \\ \dots\dots\dots\dots\dots\dots\dots &= 0, \\ \alpha_1 s + \dots\dots\dots + \alpha_4 s''' &= 0, \end{aligned}$$

must be coaxal; by effecting a linear transformation of the α_i we may take four of these points α as vertices of the tetrahedron of reference, in which case the four planes p, q, r, s are coaxal; similarly for the four planes p', q', r', s', etc.

The eight lines b arise from such values of the λ_i as make the following four planes coaxal:

$$\begin{aligned} \lambda_1 p + \lambda_2 q + \lambda_3 r + \lambda_4 s &= 0, \\ \dots\dots\dots\dots\dots\dots\dots &= 0, \\ \dots\dots\dots\dots\dots\dots\dots &= 0, \\ \lambda_1 p''' + \dots\dots\dots\dots\dots &= 0. \end{aligned}$$

It is clear that any line b must meet each of the four preceding lines a except e.g. when the λ_i are such that

$$\lambda_1 p + \lambda_2 q + \lambda_3 r + \lambda_4 s \equiv 0;$$

and there cannot be more than two such identities, for in that case the four planes p, q, r, s would coincide, and Δ would break

* F. Schur, *Ueber eine besondre Classe von Flächen vierter Ordnung*, Math. Ann. xx.

up into factors. Hence it follows that the line (p, q, r, s) must meet at least six lines b; so that if the lines (p, q, r, s), (p', q', r', s') intersect there must be at least four lines b which meet each of them, which is impossible since the order of Δ is four.

Hence no two lines a can intersect; it follows that the lines a are different from the lines b. Moreover a line a cannot meet more than six lines b, for suppose it meets seven lines b, then since any three lines a would meet a common set of five lines b, the quadric through these eight lines would meet any k_6 in fifteen points, since every b_i is a trisecant of every k_6, and hence would contain it. Therefore any line a meets exactly six lines b; similarly any line b meets exactly six lines a. In a case in which a line b does not meet a line a, e.g. (p, q, r, s), an identity exists of the form

$$\Lambda_1 p + \Lambda_2 q + \Lambda_3 r + \Lambda_4 s \equiv 0,$$

so that p may be replaced by zero in equations (1), Art. 97.

Take therefore four lines $b_1 \dots b_4$ such that each of them does not meet *two* of the lines $a_1 \dots a_4$, e.g. a_2 and a_3, a_3 and a_4, a_4 and a_1, a_1 and a_2 respectively, then equations (2), Art. 97, may be reduced to the form

$$\begin{aligned} &\alpha_1 p + \dots\dots\dots\dots + \alpha_4 p''' = 0, \\ &\alpha_1 q + \alpha_2 q' \dots\dots\dots\dots\dots\dots = 0, \\ &\dots\dots \alpha_2 r' + \alpha_3 r'' \dots\dots\dots = 0, \\ &\dots\dots\dots\dots\dots \alpha_3 s'' + \alpha_4 s''' = 0. \end{aligned}$$

The required surface is therefore

$$pq'r''s''' = p'''qr's''.$$

This surface being susceptible of collineation into itself, if it be represented by $\Delta + \Delta' = 0$, then *either* Δ and Δ' are interchanged by the collineation, *or* the planes which constitute Δ are cyclically permuted: similarly for Δ'. An instance of the former is given by the surface

$$16x_1x_2x_3x_4 + (x_1 + x_2 + x_3 + x_4)(x_1 + x_2 - x_3 - x_4) \\ \times (x_1 - x_2 + x_3 - x_4)(x_1 - x_2 - x_3 + x_4) = 0;$$

if this be written in the form

$$16x_1x_2x_3x_4 + a_x b_x c_x d_x = 0,$$

it is seen to be unaltered by the collineation

$$x_1 : x_2 : x_3 : x_4 = a_y : b_y : c_y : d_y.$$

The surface is desmic.

In the latter case, viz. when the faces of Δ are cyclically permuted by the collineation, the latter must be of period four; taking Δ as tetrahedron of reference the collineation is then of the form

$$\rho x_1 = k_1 y_2, \quad \rho x_2 = k_2 y_3, \quad \rho x_3 = k_3 y_4, \quad \rho x_4 = k_4 y_1.$$

By a change of the coordinate system we may take each k_i to be unity.

The equation of the surface is now seen to be

$$\Delta + \Delta' \equiv K x_1 x_2 x_3 x_4 + \alpha_1 \alpha_2 \alpha_3 \alpha_4 = 0,$$

where

$$\alpha_1 \equiv u_1 x_1 + u_2 x_2 + u_3 x_3 + u_4 x_4,$$

$$\alpha_2 \equiv u_2 x_1 + u_3 x_2 + u_4 x_3 + u_1 x_4,$$

$$\alpha_3 \equiv u_3 x_1 + u_4 x_2 + u_1 x_3 + u_2 x_4,$$

$$\alpha_4 \equiv u_4 x_1 + u_1 x_2 + u_2 x_3 + u_3 x_4.$$

Conjugate tetrahedra.

Denoting by Φ_1 the quadric $\sum_1^4 u_i S_i = 0$, where

$$S_1 \equiv x_1^2 + x_3^2 + 2x_2 x_4, \qquad S_2 \equiv 2(x_1 x_2 + x_3 x_4),$$

$$S_3 \equiv x_2^2 + x_4^2 + 2x_1 x_3, \qquad S_4 \equiv 2(x_1 x_4 + x_2 x_3),$$

it is clear that the planes $\alpha_1 \ldots \alpha_4$ are the polar planes of the vertices $A_1 \ldots A_4$ of Δ for the quadric Φ_1. Two tetrahedra such that the faces of one are the polar planes for a quadric of the vertices of the other, may be termed *conjugate*. Again it is easily seen that Δ and Δ' are conjugate for the quadric

$$\Phi_2 \equiv u_1 S_2 + u_2 S_3 + u_3 S_4 + u_4 S_1 = 0,$$

and hence for each of the quadrics

$$\Phi_3 \equiv u_1 S_3 + u_2 S_4 + u_3 S_1 + u_4 S_2 = 0,$$

$$\Phi_4 \equiv u_1 S_4 + u_2 S_1 + u_3 S_2 + u_4 S_3 = 0;$$

since Φ_3, Φ_4 are the quadrics obtained by submitting Φ_1 and Φ_2 to the given collineation.

Hence Δ and Δ' are conjugate in four ways.

Now it can be shown that when the tetrahedra Δ, Δ' are conjugate, four faces of Δ meet four faces of Δ' in four lines which belong to the same regulus* of a quadric.

Now since Δ, Δ' are conjugate with regard to each of four quadrics, it occurs four times that four intersections of their faces are co-regular. But if a quadric Σ meet a non-ruled quartic surface F^4 in four lines of a regulus, it will meet F^4 in four other lines of the complementary regulus; since if c_4 be this residual curve of intersection, from each point of c_4 a line can be drawn to meet the four given lines, this line therefore lies upon F^4, hence c_4 must consist of four lines of the other regulus of Σ. Therefore corresponding to each way in which Δ, Δ' have four intersections co-regular we obtain four lines of Δ, giving, in addition to the sixteen lines of intersection of Δ and Δ', sixteen other lines upon the surface.

The existence of these thirty-two lines upon the surface may also be seen from the expression of the surface in the form

$$pq'r''s''' = p'''qr's'';$$

for this shows the existence of eight lines not included in the eight lines a or the eight lines b, e.g. $p = r' = 0$, etc.

If we had started with the other four lines a and b we should have obtained a second form of the equation of the surface in the form

$$\Delta_1 + \Delta_1' = 0,$$

where Δ_1, Δ_1' are two new tetrahedra; they again yield eight lines not included in the eight lines a and b.

* For Δ being the tetrahedron of reference, and the quadric with regard to which Δ and Δ' are conjugate being $\Sigma a_{ik} x_i x_k = 0$, the four lines just referred to are

$$x_1 = 0, \quad a_{12}x_2 + a_{13}x_3 + a_{14}x_4 = 0, \text{ etc.}$$

If the join of two points X, Y meets this line we have

$$\frac{X_1}{Y_1} = \frac{a_{12}X_2 + a_{13}X_3 + a_{14}X_4}{a_{12}Y_2 + a_{13}Y_3 + a_{14}Y_4},$$

hence if $p_{ik} = X_iY_k - X_kY_i$, it follows that

$$p_{12}a_{12} + p_{13}a_{13} + p_{14}a_{14} = 0.$$

The conditions that p_{ik} should meet the other three lines are seen to be, similarly,

$$p_{21}a_{21} + p_{23}a_{23} + p_{24}a_{24} = 0,$$
$$p_{31}a_{31} + p_{32}a_{32} + p_{34}a_{34} = 0,$$
$$p_{41}a_{41} + p_{42}a_{42} + p_{43}a_{43} = 0;$$

and since $p_{ik} = -p_{ki}$, $a_{ik} = a_{ki}$, the four equations are equivalent to three; hence an infinite number of lines p meet the four given lines which therefore belong to the same regulus.

118. Tetrahedra subject to two collineations.

We now consider the case in which a collineation of period four permutes the faces of Δ, and a collineation of period three permutes three of these faces and leaves the fourth unaltered; the tetrahedron Δ' being similarly affected.

Taking the collineations as

$$\rho x_1 = y_2, \quad \rho x_2 = y_3, \quad \rho x_3 = y_4, \quad \rho x_4 = y_1 \quad \text{........(I)},$$

$$\sigma x_1 = y_1, \quad \sigma x_2 = y_3, \quad \sigma x_3 = y_4, \quad \sigma x_4 = y_2 \quad \text{........(II)},$$

it is seen that the surface

$$Ax_1x_2x_3x_4 + (mx_1 + x_2 + x_3 + x_4)(x_1 + mx_2 + x_3 + x_4)(\ldots)(\ldots) = 0$$

is unaffected by each collineation.

There are six planes each containing two intersections of faces of Δ and Δ' and two other lines, viz. the planes

$$x_1 + x_2 + m(x_3 + x_4) = 0.$$

The surface therefore possesses $16 + 12 = 28$ lines; it has six coplanar nodes, viz. the points $x_1 = x_2 = \sum_1^4 x_i = 0$, etc.

Next consider the collineation

$$\rho x_1 = y_2, \quad \rho x_2 = -y_3, \quad \rho x_3 = y_4, \quad \rho x_4 = y_1 \quad \text{.....(III)},$$

the collineation (II) being as before.

The surface

$$Ax_1x_2x_3x_4 + (x_2 + x_3 + x_4)(x_1 - x_2 + x_3)(x_1 - x_3 + x_4)(x_1 + x_2 - x_4) = 0$$

is unaffected by the collineations (II) and (III).

The tetrahedra Δ, Δ' are conjugate in nine ways; in six ways arising from the six quadrics

$$x_3^2 - x_4^2 + 2x_1(x_3 + x_4) - 2x_2(x_3 - x_4) + 2x_1x_2 = 0, \text{ etc.},$$

and in three ways arising from the three quadrics

$$x_1^2 - x_2^2 + x_3^2 - x_4^2 + 2x_1x_2 + 2x_3x_4 + 2x_2x_3 - 2x_1x_4 = 0, \text{ etc.};$$

each manner in which Δ and Δ' are conjugate gives rise to four lines on the surface, which is thus seen to possess $16 + 9 \times 4 = 52$ lines in all. Each of the tetrahedra Δ, Δ' is inscribed in the other.

INDEX OF SUBJECTS

The numbers refer to the pages

Anchor-ring 107
Apolar conics 136; quadrics 182
Asyzygetic surfaces 10

Base-points of representative cubics 41
Bauer's surfaces 189
Bidouble line 70, 78
Biplanar nodes of three species 67

Class of surface with double conic 46
Close-points 70
Coincident base-points 64
Collineations which leave two tetrahedra unaltered 193, 195
Cones of first species in four dimensions 56; of second species 77
Confocal cyclides 109, 115
Conics on surface $\Phi^2=\alpha\beta\gamma\delta$ 133; on surface with double conic 38; on surface with double line 119, 130
Conjugate tetrahedra 193
Constants, number of in quartic surface with a double conic 39
Contact-conic 3
Correspondence of points upon the surfaces Δ 163; the Jacobian of four quadrics 166
Cremona transformation 156
Cubic complex connected with desmic surface 27
Cubic curves on surface with double conic 43; with double line 124
Cuspidal double conic 69
Cyclide, canonical forms of equation 99; Cartesian equation of 86; conjugate points on 115; shapes of 91, 101; lines of curvature of 117
Cyclides, confocal 109, 115, 116; corresponding points on 111

Desmic surface 26; conics on 37; plane sections of 31–35; quartics on 27; section by tangent plane 35; sixteen points, group of 32
Desmic tetrahedra 24
Determinant surfaces 161; correspondence of points 163; sextic curves on 162
Dianodal surface 10; curve 12
Double plane 152
Dupin's cyclide 106

Focal curves of cyclide 87, 91
Fundamental inversions of quartic surface with double conic 59
Fundamental quintic for cyclide 85, 93

Hessian of cubic surface 190

Inverse of cyclide 107
Inverse points on cyclide 89
Involutory property of the surface $U^2+x_1{}^3x_2=0$ 70

Jacobian of four quadrics 2, 165; cubics on 169; quartics on 171; sextics on 171; ten lines of 167; tangent plane of 169

Kummer's surface 22, 173, 184

Monoid, the quartic 147; twelve lines of 147; with six nodes 149; cubics and quartics of 147

Nodal sextic curves 4
Nodes, maximum number of 1
Normals to cyclide from external point 108

Orthogonal system of five spheres 97

Pentaspherical coordinates 98
Perspective relation of surface with double conic and general cubic surface 49
Pinch-points on surface with double conic 38

Plücker's surface 128
Power of two spheres 94
Principal spheres of a cyclide 87

Quadrics, five associated with cyclide 87; inscribed in surface with double conic 59, 79
Quartic curves on surface with double conic 38, 45, 58; with double line 125
Quartic, plane, and surface with double conic 52

Rational surfaces 152
Rationalization of surface with double conic 40; with double line 120; Plücker's surface 128; Steiner's surface 135
Ruled surfaces, arising in special case 68

Schur's surfaces 191
Sextic curves, plane, with ten nodes 6; with seven nodes 4
Sigma functions, in desmic surfaces 27
Sixteen lines of surface with double conic 39, 46; with double line 118
Sphero-conics on cyclide 113
Steiner's surface 81, 132, 134; asymptotic lines of 137; conics on 35; quartics on 136, 141
Symmetroid 14, 166; ten nodes of 167; sextic curves on 171; curves of ninth order 170; curves of fourteenth order 171
Syzygetic surfaces 10

Table of forms of surfaces with double conic 82–85
Tacnode 132, 152
Ten lines of Jacobian of four quadrics 167
Ten-nodal sextic curves 6
Torsal line 37, 63, 127
Trisecants of sextic curves on surfaces Δ 164
Trope 4

Uniplanar node 68

Weddle's surface 173; asymptotic lines of 177; conjugate quintics 176; connection with Kummer's surface 184; expression of coordinates as double Theta functions 185; forms of equation 178; group of thirty-two points 180; octavic curves on 185; plane section of 189; septic curve on 177

INDEX OF AUTHORS

The numbers refer to the pages

Baker 130, 169, 181, 183, 187
Bateman 175, 183
Bela Totössy 70
Berry 134
Bioche 37
Bobek 48, 51
Bôcher 99, 103
Bromwich 72, 99

Caspary 185
Cayley 10, 166, 167, 168, 169
Clebsch 34, 41, 120, 135, 152
Conner 189
Cotty 137
Cremona 35, 49

Darboux 86, 108
de Franchis 134

Eckhardt 145

Geiser 46, 49, 182

Harkness 29, 31
Hesse 152
Hierholzer 178
Hudson 183, 186, 187
Humbert 24, 32, 113, 116, 187

Korndörfer 64
Kummer 38, 131

Lachlan 94
Lacour 137

Morley 29, 31, 189

Nöther 133, 152, 157

Reye 139
Richmond 174, 184
Rohn 3, 147

Schröter 143
Schur 163, 191
Segre 55, 72, 78, 82
Serret 183
Sisam 133
Sturm 139, 140

Weierstrass 143
Weiler 49

Zeuthen 52